The Traumatic Stress Recovery

WORKBOOK

40 Brain-Changing Techniques You Can Use Right Now to Treat Symptoms of PTSD & Start Feeling Better

Jennifer Sweeton, PsyD

New Harbinger Publications, Inc.

Publisher's Note

This publication is designed to provide accurate and authoritative information in regard to the subject matter covered. It is sold with the understanding that the publisher is not engaged in rendering psychological, financial, legal, or other professional services. If expert assistance or counseling is needed, the services of a competent professional should be sought.

NEW HARBINGER PUBLICATIONS is a registered trademark of New Harbinger Publications, Inc.

New Harbinger Publications is an employee-owned company.

Library of Congress Cataloging-in-Publication Data

Names: Sweeton, Jennifer, author.
Title: The traumatic stress recovery workbook : 40 brain-changing techniques you can use right now to treat symptoms of PTSD and start feeling better / Jennifer Sweeton.
Description: Oakland, CA : New Harbinger Publications, [2022] | Includes bibliographical references.
Identifiers: LCCN 2022018876 | ISBN 9781684038848 (trade paperback)
Subjects: LCSH: Post-traumatic stress disorder--Treatment. | Self-care, Health. | BISAC: SELF-HELP / Post-Traumatic Stress Disorder (PTSD) | PSYCHOLOGY / Psychopathology / Post-Traumatic Stress Disorder (PTSD)
Classification: LCC RC552.P67 S94 2022 | DDC 616.85/21--dc23/eng/20220716
LC record available at https://lccn.loc.gov/2022018876

Printed in the United States of America

24   23   22

10   9   8   7   6   5   4   3   2   1   First Printing

"*The Traumatic Stress Recovery Workbook* delivers practical and accessible techniques for readers with diverse trauma backgrounds to approach the recovery process with confidence. Jennifer Sweeton's years of professional experience and deep understanding of effective trauma treatment are on full display as each chapter masterfully balances psychoeducation, skill building, and personal reflection. This workbook will be a game-changer for countless people impacted by trauma."

> —**Corey Petersen, PhD**, clinical psychotherapist, and
> psychotherapeutic language researcher and educator

"This book puts together an exceptional understanding of how trauma impacts the brain and body. This collection of tools provides individuals who have experienced trauma with the means to combat the associated symptoms and impact on their lives. This workbook has something for anyone who has ever struggled."

> —**Amanda S. Vaught, PsyD**, trauma expert,
> former director of training within Veteran Affairs Health Care System,
> and adjunct faculty at Villanova University

This book is dedicated to my precious daughter,
Annaliese, who inspires me to try new things,
work hard, and live life to the fullest!

Contents

Introduction

If you've experienced psychological trauma, you're probably aware of how it can impact your relationships, professional life, and overall functioning. In fact, many survivors report that there is no domain of their life that it hasn't touched in some way. However, if this has happened to you, you're not alone, and recovery *is* possible! As a trauma psychologist, I've treated hundreds of people who have been traumatized by war, abuse, neglect, natural disasters, accidents, sexual assault, and various other forms of violence. Despite all of the pain and heartbreak I've heard about, I remain hopeful and optimistic about everyone's potential to heal, as I've seen amazing growth and recovery even after the most debilitating trauma.

While everyone's story is different, and no two people's healing journeys are alike, I've found that people's reactions to trauma can actually be quite similar, and many clients report the same challenges after trauma. For instance, a couple of common reactions to trauma are sleep difficulties and anger problems, and the good news is that there are techniques that can help you address these issues. This book describes nine of the most common challenges facing trauma survivors and offers forty tools and techniques you can begin using right away. It is my sincere hope that you will find this workbook helpful in your recovery journey!

Self-Assessment: Is This Book for You?

As a unique person with your own reactions, there is no research study or book that can fully capture and address the entirety of your experience. However, this book may be for you if some of the following challenges seem familiar to you. For example, if you've found it difficult to trust other people, or manage anger after your trauma(s), you're not alone, and this book may be beneficial as you learn how to tackle these challenges. Take a look at the self-assessment below, and ask yourself whether these topics, or themes, resonate with you. You can also visit http://www.newharbinger.com/48848 to download a copy of this exercise.

Common Challenges After Trauma

Common Challenge	Have I Experienced This? (yes/no)	If So, How Problematic or Distressing Is It? (1–10, 1 = not at all, 10 = extremely)
Intrusive thoughts about the trauma		
Feeling anxious and revved up		
Difficulties with anger		
Strong, distressing physical reactions		
Emotional numbing and apathy		
Issues around safety and control		
Difficulties with trust and relationships		
Isolation and agoraphobia (fear of leaving home)		
Sleeping difficulties		

Reviewing your answers, what do you notice? Did you answer "yes" to any of the topics in the Common Challenge column? If you did, and if you scored a 5 or above on one or more of the challenges, then this book was written for you. Like many other trauma survivors, you're likely experiencing post-trauma difficulties that are impacting your life. The tools and exercises in this book help survivors begin to address these exact challenges.

How to Use This Book

While it may work fine for you to start at the beginning of this book and read it through to the end, that may not be what you're most needing. Instead, begin by reading this introduction in its entirety, in order to understand the neuroscience and types of techniques referred to throughout the book. The next two sections of this introduction cover these two topics. Next, read the chapters that focus on the topics you find to be most problematic or challenging for you. Then, as is relevant to you, or desired, read the remaining chapters. Each chapter is designed to stand alone, though some chapters reference techniques described in other chapters, which you can easily locate. Note: There are many exercises and worksheets in this book that you can also download at this book's website: http://www.newharbinger.com/48848. (See the very back of this book for more details.)

The Neuroscience of Trauma

Although this book frequently references the neuroscience of trauma, don't worry—you don't need to become a neuroscientist to understand it! The brain is immensely complex, but some of the main brain areas involved in psychological trauma can be summed up in a straightforward manner. Here's a quick introduction to the four main areas of the brain impacted by trauma:

1. **The amygdala, known as the "smoke detector" of the brain:** This tiny, almond-shaped brain structure lies deep inside your brain. Though it be tiny, it is very powerful, as it is the area of the brain that alerts your entire brain and body when danger, or a threat, may be present. It is called the "smoke detector" because its main job is to detect signs of danger, even small ones. When the amygdala senses that something is wrong, and potentially harmful, it activates, sending alarm signals to your whole body, instructing it to prepare to fight, flee, or freeze as a way of coping with the threat. When this occurs, you'll definitely feel it! Amygdala activation feels like stress, upset, anxiety, sadness, and even rage. You may know the amygdala has been activated when you feel frantic, a racing heart, shallow breathing, muscle tension, or other physical symptoms of stress. After trauma, it is common for the amygdala to activate more easily, even in response to nonthreatening situations. It can also occur that the amygdala "switches on" after trauma, so the person

feels on edge, vigilant, and stressed *all* the time, regardless of the circumstances. A main goal of trauma treatment is to lower activation of this brain area, which will allow you to think more clearly and feel calmer.

2. **The insula, known as the "body sensor" of the brain:** This brain structure is extremely important, as it helps you experience and understand your emotions. Specifically, the insula alerts you to sensations occurring in your body (such as rapid heart rate or muscle tension), and then helps you interpret, or understand, what those sensations mean with regard to emotion. For example, a racing heart could be interpreted by the insula as fear or anxiety ("I am scared of roller coasters!") or it could signal excitement ("Roller coasters are thrilling!"). How you experience your body, and which emotions are derived from your sensations, is largely dependent on your insula. After trauma, the insula's activation can lower, making it difficult to connect with your body and your emotions. This can lead to numbing, or apathy. Thus, after trauma, it is important to strengthen this area of the brain.

3. **The hippocampus, known as the "memory storage unit" of the brain:** A lot of brain regions are involved in memory, but the hippocampus is arguably the most well-known. In particular, the hippocampus, which is located close to the amygdala, helps you encode and store long-term memories. Whenever you remember something that happened in the past, you're pulling that information from the hippocampus. After trauma, the hippocampus can become less activated, and can actually shrink, due to cortisol (the stress hormone) damaging it. In other words, the stress of trauma can damage this area of the brain. Thankfully, however, the hippocampus can be strengthened and reactivated, which is a goal of many trauma-focused therapies.

4. **The prefrontal cortex, known as the "thinking center" of the brain:** This large brain region is near the top of your brain, located behind your forehead. It's known as the "thinking" or "executive functioning" center of the brain because of its involvement in planning, goal setting, logical thinking, reasoning, and concentration. When your prefrontal cortex is working well and is adequately activated, you can reason through and evaluate situations well; you are clearheaded. After traumatic events, however, this region can become less activated, making it difficult to focus, make decisions, regulate emotions, or make sense of the world. Thus, it is

important for trauma survivors to practice skills and techniques that strengthen the prefrontal cortex.

Bottom-Up versus Top-Down Techniques

The skills in this workbook will help you facilitate your own brain change after trauma, altering the amygdala, insula, hippocampus, and prefrontal cortex in the direction of health. In each chapter, you'll read a bit about which post-trauma challenges your brain needs to address and several techniques will be presented for you to try. These techniques, or skills, tend to fall into three categories: behavioral techniques, bottom-up techniques, and top-down techniques.

- **Behavioral techniques:** These techniques emphasize the importance of behavior. While other techniques may ask you to think about something or notice something about your experience, behavioral techniques focus on what you *do*, or how you behave. Sometimes, the best way to think or feel differently is to actually act differently. For example, a basic concept in exposure therapy is that if you want to conquer something that you fear, you have to face it, or do it. Some exercises in this workbook encourage you to engage in certain healthy behaviors, which can help rewire and shift the brain areas described earlier, making you feel more confident and less stressed.

- **Bottom-up techniques:** These techniques work with the body, including physical sensations and the breath, to change the brain after trauma. Some bottom-up techniques include paying attention to the body, engaging in deep-breathing exercises, or moving the body in particular ways (as you would in yoga). Because the brain is connected to the body, a great way to change the brain is to work with the body, and that's exactly what bottom-up techniques help you do. You'll notice that some bottom-up techniques are recommended in every chapter of this book, and for good reason, because bottom-up techniques are often especially effective at helping you learn how to decrease amygdala activation (the brain's smoke detector) and increase insula activation (the brain's body sensor). Some of these techniques can also strengthen the hippocampus (the brain's memory storage unit) and prefrontal cortex (the brain's thinking center).

- **Top-down techniques:** These techniques can also facilitate healthy brain change, but in a different way than bottom-up techniques. While bottom-up techniques work with the body, from the "bottom up," to change the brain, top-down techniques work with your mind, or thoughts, to change the brain "top down." Top-down exercises include analyzing scenarios, seeing whether there are other ways of interpreting situations, thinking differently about your life, and some types of meditations. These skills can be especially helpful when you want to increase the prefrontal cortex's activation or strengthen the hippocampus, although some techniques can also positively impact the insula and amygdala.

Although a lot of techniques can be considered hybrid skills, containing both behavioral *and* bottom-up elements, or bottom-up *and* top-down elements, there is usually one category that best describes any given technique. In this workbook, you'll notice that techniques are noted as being in one or more of these categories.

CHAPTER 1

Intrusive Thoughts About the Trauma

Deshaun, a thirty-year-old man, stopped at a gas station for a soda one night two years ago. While he was inside, an armed robber burst into the gas station and held everyone inside hostage. While Deshaun was not physically hurt during this event, he witnessed the shooting of a gas station attendant and two customers. Since that time, no matter how hard he has tried to put the event out of his mind, he finds himself replaying some of the details of that night over and over, almost every day, and even has nightmares about it at night.

Check-In: Your Experiences

After reading about Deshaun, take a moment to think about your own experiences. You may choose to reflect and journal on the following questions:

- Have you ever felt similar to Deshaun, where you have replayed something terrible that happened over and over?

- Have you tried hard not to think about something traumatic that occurred? When you push it out of your mind, what happens? What, if anything, tends to help you when you try to avoid thoughts and feelings about the event?

- How have intrusive thoughts impacted your life? What was life like before you had to manage these thoughts?

How Intrusive Thoughts May Impact Your Life

Relentless, intrusive thoughts can be frustrating, scary, and exhausting to manage, and are unfortunately very common after traumatic events. If you struggle with intrusive thoughts, you may notice that the thoughts negatively impact multiple domains of your life, including family, friendships, work, school, and productivity more broadly. In particular, it is not unusual for trauma survivors to appear distracted and disengaged from family and friends as a result of their mind getting stuck on traumatic memories from the past. When this happens to you it can feel like a broken record, or a loop, where you replay a piece of your history over and over. Additionally, when unwanted thoughts take over, this can impact performance at school or work, because these thoughts make it extremely difficult to focus on the task at hand. You may find that you have to read the same material over and over to digest it, sometimes forget tasks, and easily lose your train of thought, as your mind seems constantly drawn to thoughts about the trauma(s). Over time, this can lead to exhaustion and increased anxiety and depression, in addition to more severe post-traumatic stress disorder (PTSD) symptoms.

The following brief self-assessment is a fast way to determine whether you might be experiencing problematic, unwanted, and intrusive trauma-based thoughts. You can also visit http://www.newharbinger.com/48848 to download a copy of this exercise.

Problematic Intrusive Thoughts Self-Assessment

Read the following and circle the best response on a scale of 0 to 3, where 0 = None/not at all, 1 = A little bit/sometimes, 2 = A moderate amount/often, and 3 = A lot/most of the time.

Since a trauma (or multiple traumas) happened, I find that I think about it a lot.	0	1	2	3
It is hard for me to stop thinking about my trauma(s) when I start thinking about it.	0	1	2	3
I think about my trauma(s) by accident, without meaning to.	0	1	2	3
I catch myself thinking about my trauma(s) without knowing I was thinking about it.	0	1	2	3
Images of the traumatic event(s) come to mind when I don't want them to.	0	1	2	3
Sounds of the traumatic event(s) come to mind when I don't want them to.	0	1	2	3
Sensations related to the traumatic event(s) come to mind when I don't want them to.	0	1	2	3
Thoughts of the trauma(s) distract me from things I need to be doing or focusing on.	0	1	2	3
Thoughts about the trauma(s) have negatively impacted my school or work productivity.	0	1	2	3
I take longer to complete tasks because I'm distracted by thoughts of the trauma(s).	0	1	2	3
It is distressing to me when I think about the trauma(s).	0	1	2	3
My body reacts with distress (racing heart rate, stomach pain, chest tightening, etc.) when I think about the trauma(s).	0	1	2	3
I feel sad, angry, scared, or other distressing emotions when I think about the trauma(s).	0	1	2	3
The strategies I've tried to stop thinking about the trauma(s) do not work well.	0	1	2	3
I have tried drinking alcohol or doing drugs to stop thinking about the trauma(s).	0	1	2	3
Other people have told me I seem distracted or distant from them when I've been thinking about the trauma(s).	0	1	2	3
When I think about the trauma(s), it puts me in a bad mood.	0	1	2	3
My thoughts about the trauma(s) negatively impact my relationships with other people.	0	1	2	3
My thoughts about the trauma(s) feel like a movie reel playing over and over, where it is difficult to stop the reel from playing.	0	1	2	3
Even small "triggers" make me start to think about the trauma(s).	0	1	2	3
Thoughts of the trauma(s) keep me up at night, making it hard to fall or stay asleep.	0	1	2	3
Total				

Add up the total score from the above items, and record it next to "Total." A score of 24 or higher indicates the you likely experience problematic intrusive thoughts related to past trauma.

While for some people these thoughts fade within a few months of the trauma occurring, for others they may persist for many years, or indefinitely. If your intrusive thoughts are persistent, exhausting, and impacting your work or social life, seek help to better manage and reduce them. The information and exercises in this chapter can help you as you begin to address intrusive thoughts, but for many people, professional help from a licensed psychotherapist is also immensely helpful. Use the following exercise to reflect on your self-assessment score. You can also visit http://www.newharbinger.com/48848 to download a copy of this exercise.

Reflect on Your Self-Assessment Score

What did you score on the self-assessment? Was this surprising to you? Take a moment to reflect on, and process, what you just learned about yourself. You may consider the following questions as you reflect on your experience with this.

Was your score higher than you thought it would be? Lower?

_____

_____

What does your score indicate about you and the challenges you're experiencing?

_____

_____

_____

Were there items on the assessment you hadn't considered before?

_____

_____

_____

Were there items on the self-assessment that surprised you?

_____

_____

_____

Are your intrusive thoughts something you think you should address?

_____

_____

_____

Why Intrusive Thoughts After Trauma Are So Common

Intrusive, trauma-related thoughts are one of the more common, and distressing, symptoms of PTSD. While you may want nothing more than to forget about a traumatic event after it happens, your brain will likely ensure that the memory of it is never far. Though frustrating, disruptive, and often scary, intrusive thoughts persist in large part because of how traumatic memories are stored in the brain. Specifically, during moments of extreme stress or trauma, the brain encodes memories in a way that ensures they will be replayed, or thought of, often.

At the time of a traumatic event, your body and brain become flooded with cortisol (the stress hormone) as your stress response increases. When this happens, the hippocampus, the long-term memory storage unit that helps you encode memories, becomes saturated with cortisol. This is due to the large number of cortisol receptors covering the hippocampus. When cortisol rushes to the hippocampus, it can impact how the hippocampus activates and, ultimately, how trauma memories become encoded in the brain. Because cortisol can compromise the functioning of the hippocampus, traumatic memories are often encoded and consolidated differently from nontraumatic memories.

For example, when cortisol prevents the hippocampus from fully activating, a trauma may be encoded in a fragmented manner, where the sensory details and facts about a memory are stored in a disjointed way. When this happens, you may remember parts of the memory in flashes, smells, or sounds, but these experiences may feel disconnected from one another, almost like pieces of a puzzle that you struggle to put together. Fragmented memories such as these are especially difficult to cope with, because these fragments lack context or coherence that would allow you to rationally analyze them and soothe yourself. For example, if one memory fragment is of a knife, and it does not contain context ("A knife is dangerous when an intruder is pointing one at me, but not when I am eating dinner with my family…"), it becomes universally threatening and there is a generalization of fear with regard to these fragments ("The presence of a knife means my life is at risk").

Also, when a memory is shattered into a thousand pieces, like a broken mirror that you cannot reconcile or integrate, it can feel as though you're constantly stepping on shards of glass with no warning. Even small, innocuous stimuli may trigger one or more of the shards, which is less likely to happen when memories are more integrated and contain context. Thus, one thing that makes trauma memories so hard to manage is that their fragments frequently pop up, often unexpectedly. This can feel very exhausting and can lead to avoiding people, places, or situations where you fear you might become triggered.

The way trauma memories are encoded can also lead to a generalization of fear of many different stimuli, situations, people, and places, and it can make it scary to live a full life, as you may always be questioning what might trigger one of the memory shards. To better understand why intrusive trauma memories are so common, here is a summary of what happens in the brain when you experience trauma.

1. **When trauma happens, the amygdala activates.** When you experience trauma, your amygdala, the fear center of your brain that detects threat, becomes activated. This sets off a chain reaction where the rest of the brain is warned about the threat, and the threat signals are communicated down to the body. This is likely what happened when Deshaun first realized something terrible was happening in the gas station.

2. **Your body activates the stress response.** When the amygdala detects danger, it communicates with the hypothalamus in order to activate the stress pathway, called the hypothalamus-pituitary-adrenal (HPA) axis. When the HPA axis activates, the entire body enters the stress response, which prompts more than 1,400 biochemical and psychophysiological reactions, including the release of cortisol, which is often referred to as the "stress hormone."

3. **Cortisol floods the hippocampus.** Because the hippocampus (the memory center of the brain) is covered in cortisol receptors, cortisol rushes to the hippocampus and saturates it when you become very stressed or traumatized. At the same time this happens, the hippocampus begins the process of encoding the memory of what's happening during the trauma, to ensure you don't forget it. However, with cortisol flooding this brain region, it makes it difficult for the hippocampus to function as well as usual, and it may not be able to fully activate as it normally would during the memory-encoding process. This leads to traumatic memories being consolidated and encoded differently from nontrauma memories.

4. **Trauma memories are fragmented.** Because the hippocampus may not be able to fully activate at the time of memory encoding, trauma memories can be stored in the brain in a fragmented manner, where the facts, events, and sensory experiences associated with the memory are disconnected. This can lead to generalization of fear, where you develop a fear reaction to any stimulus that has *anything* in common with any of the memory fragments that were encoded ("All men are dangerous," or "I am afraid of the dark," or "All blue cars are threatening"). Also, trauma memories often contain gaps, making it difficult to remember exactly what, when, and in what order events occurred.

5. **Fragmented memories are easily triggered.** When trauma memories lack context and contain many fragments, several of which may resemble everyday life situations and objects (knives, cars, animals, etc.), it can lead to you feeling unexpectedly triggered by a variety of benign situations or stimuli. When trauma reminders are all around you, it makes it impossible to stop thinking about the traumatic event, and you'll likely notice that intrusive thoughts about the trauma occur frequently. While reminding you of the trauma is your brain's way of trying to keep you safe and on guard for danger, it can feel frustrating and overwhelming!

Use the following exercise to reflect on your trauma triggers. You can also visit http://www .newharbinger.com/48848 to download a copy of this exercise.

Your Trauma Triggers

Most of us have triggers that bring up thoughts about past traumas. Understanding what these are, and what they mean, can help you better manage them when they arise. Complete the following to learn more about your own triggers.

Times of the Day or Year That Are Triggering: You may find, over time, that certain times of the year, seasons, or even times of day feel more triggering than others, prompting intrusive trauma-related thoughts. This may be due to weather changes, the presence of sunlight/darkness, smells that are associated with different times of the year (such as holidays), and so on. Are there times of the year, week, or day that you tend to experience increased intrusive thoughts? Can you reflect on why this might be the case?

_____

_____

_____

_____

Places That Are Triggering: There may also be places that feel triggering to you, where, when you visit them, you are suddenly flooded with trauma memories. This is probably especially true if you experienced a traumatic event in a location you frequently have to go to or near. Are there specific places or broad regions that seem to elicit traumatic memories? What is it about these places that is reminiscent of the trauma?

_____

_____

_____

_____

People Who Are Triggering: People can be very strong triggers, as they can bring up old thoughts, feelings, and beliefs associated with your trauma. Have you noticed any specific people who, when you're around them, tend to remind you of your trauma(s)? This may even occur with people who were not involved in the trauma(s). If you can identifying triggering people, what is it about them that might be leading your brain to remind you of your past traumas?

_____

_____

_____

_____

_____

Situations That Are Triggering: Sometimes it's not just one place, or one time of year, or one person who triggers intrusive trauma thoughts. Rather, it can be a combination of all of these things. For instance, you may be fine during December, but once family members arrive at the house, it begins snowing, the kitchen smells like pie, and you see holiday decorations, you start to feel haunted by past trauma. Are there "perfect storms," so to speak—combinations of people, places, and times—that tend to prompt trauma-related memories?

_____

_____

_____

_____

_____

Emotions and Sensations That Are Triggering: Most people overlook this category of triggers, but it is very important to consider. Sometimes you may find that you feel triggered, and that your mind becomes stuck on thoughts of a traumatic event, but you have no idea why. There may be several reasons for this, but one explanation is that you may have suddenly encountered a situation, or just had a thought, that has produced similar _emotions_ or *sensations* to those you experienced during a traumatic event. The situation may look totally different, and you can't identify anything in your environment that would be triggering—perhaps you just went for a jog, and while on the jog your heart began to race. If at the time of the trauma you experienced rapid heart rate, then it is possible that increased heart rate, even when unrelated to your trauma, may prompt memories of the trauma. Are there certain emotions or physical sensations that might elicit trauma memories for you?

_____

_____

_____

_____

_____

You may notice from your responses that you're prone to experiencing triggers and trauma reminders when around certain people, places, or situations, or when you experience specific sensations or emotions. Knowing this about yourself can be helpful, as you can prepare for these situations and work to learn how to manage them so that you don't have to avoid them.

Terms Your Therapist Might Use to Explain This Challenge

If your experiences are similar to Deshaun's, your therapist may use the following descriptors and terms in your discussions, or in your therapy notes:

- **Reexperiencing symptoms.** This refers to a category of PTSD symptoms and includes intrusive memories of a traumatic event. Intrusive thoughts about trauma are called "reexperiencing" because when they occur, you may feel as though you are reliving some pieces of the trauma. For example, you may start to experience similar emotions or thoughts as you did at the time of the trauma. Or you may just remember the trauma in vivid detail, making it seem like it happened very recently (or as though it's still happening currently).

- **Rumination.** This refers to a common phenomenon that is experienced in depression, anxiety, PTSD, and some other conditions. Rumination occurs when you repeatedly think about something distressing and have difficulty stopping. Sometimes this is referred to as "spinning thoughts" or "racing thoughts," and it happens when you keep replaying something over and over in your mind, usually with an intention of "solving" it or somehow making yourself feel better about it. However, usually rumination has the opposite effect; it can make you feel worse over time and the thoughts or memories may seem to never resolve.

What Your Brain Needs

If you struggle with intrusive trauma thoughts, your brain isn't broken! In fact, it's a good sign that your brain works *really* well, as it has successfully learned how to not forget horrible things that have hurt you. If you think about it, a malfunctioning brain, one that doesn't prioritize your survival, would be one that foolishly forgets the things that have threatened you (or others) before. If you struggle with intrusive memories, it means your brain is primed to learn and change, and this is great news. Here's what your brain needs if you struggle with intrusive trauma memories.

1. Bottom-up techniques that reduce the activation of the threat detection center (amygdala). The amygdala may activate in response to thoughts about a traumatic event, but it may also activate due to something totally unrelated, which may then prompt a memory of a traumatic event to come to mind. In both cases, one thing that can help you better manage (and perhaps even reduce) intrusive thoughts is to downregulate the amygdala.

2. Top-down techniques that strengthen the executive functioning and self-regulation areas of the brain (mainly the prefrontal cortex) and the memory center of the brain (hippocampus). Top-down, cognitive approaches that strengthen the brain can be immensely helpful in training your brain to react differently to trauma triggers, and to reconsolidate traumatic memories so that they feel less distressing over time. While attending psychotherapy is the best way to learn how to reconsolidate traumatic memories, you may be able to facilitate this process, to an extent, on your own.

The next section details five skills that contain both bottom-up and top-down components that can help you learn how to better manage intrusive trauma thoughts. When you are able to downregulate your amygdala and strengthen your prefrontal cortex and hippocampus, you will be able to reframe and react better to traumatic memories when they arise.

Skills for Meeting the Challenge

In this section, you'll learn and practice five skills that can help you better manage, and reduce the intensity of, intrusive traumatic memories. These skills include bilateral stimulation when you experience trauma reminders, visualization techniques, meditation, and cognitive restructuring.

Skill 1: Bilateral Stimulation with Memory Replay (Bottom-Up and Top-Down)

This exercise is categorized as a bottom-up technique with some top-down elements, where you will calm the amygdala using bilateral stimulation, while also strengthening your prefrontal cortex and hippocampus by selectively engaging with different components of a memory. Bilateral stimulation (BLS), put simply, is when you alternate movement or stimulation between the left and right sides of your body. This technique is meant to be practiced when you find it difficult to get a traumatic memory, or part of a memory, out of your mind. Importantly, do *not* intentionally try to bring a traumatic memory to mind with this technique; instead, practice it when a traumatic memory intrudes, and you cannot seem to get it out of your mind. If you notice that you're replaying part of the trauma, or experiencing trauma-focused rumination, follow these steps.

1. **Notice that you're thinking about the trauma.** This is the hardest step! You may realize that you've been thinking about something terrible or distressing long after the memory started replaying. It's not uncommon for people to get stuck in trauma-based thoughts and memories and only realize it several minutes (or even hours!) later. To get better at noticing your thoughts, answer the following:

   a. **When do you tend to think about the trauma?** You may refer to the Your Trauma Triggers exercise you completed earlier in this chapter to answer this question.

b. **When can you check in with yourself about your thoughts each day?** Sometimes the best way to begin to observe your thoughts is to set an alarm on your phone or set a calendar reminder. If you make it a point to check in with your own thinking multiple times each day, you'll get a better sense of your mind's habits.

2. **Notice your reaction to your trauma thoughts and practice self-compassion.** When you realize that your mind has wandered over to a traumatic event, notice your reaction to it. Are you hard on yourself about it? Self-blaming? Frustrated? Angry? Sometimes these reactions to yourself can happen very quickly, without you even meaning to be self-critical. If this happens, spend a brief moment shifting your approach to yourself by doing the following:

   a. **Validate your brain.** Acknowledge that your brain is reminding you of danger in an attempt to help you stay safe and alive. While frustrating, your brain is simply doing what it can to help you. As you reflect on your own brain, how might you validate what your brain is doing when it replays trauma?

   b. **Respond to your thoughts.** Once you acknowledge that your brain's intent is not to hurt you, you can gently respond by asking for a bit of space or some relief. You might say to yourself, "Brain, I know that you mean well, but we don't need these reminders right now and I'd like it if you could tone it down a little." How might you kindly ask your brain to tone it down and give you some space?

3. **Change the subject in your mind.** While "thought stopping" is not always effective, sometimes after we acknowledge and validate our own experiences, they pass through and give us some relief. To see whether this might be possible for you, gently attempt to shift your attention to another, more neutral or pleasant thought. What are some thoughts you'd like to intentionally shift toward?

4. **If trauma must replay, allow it to replay for a moment.** If step 3 was unsuccessful for you, and the thoughts and memories persist, you can then attempt to facilitate desensitization to them by using bilateral stimulation. For example, you may practice BLS by alternately tapping your left and right shoulders with each hand. Though it may sound strange, practicing *fast-paced* BLS as you think about the memory can help dull it and make it feel less upsetting. One reason for this is that the fast

tapping motion, and the sensations you feel, serve as a bit of a distraction from the memory, and this facilitates desensitization to the distressing material. Here are some ideas for incorporating fast-paced BLS when you cannot stop a trauma memory from playing:

a. Quickly tap each knee with your hands, alternating left, right, left, right.

b. Quickly tap each foot, alternating left, right, left, right.

c. Quickly tap each shoulder with your hand, alternating left, right, left, right.

d. Go for a quick-paced walk.

Integrating fast BLS allows the trauma memory to play in your mind for a few moments. For a lot of people, the memory will more easily pass once it has "played out" in the mind a bit. After a few minutes of allowing the memory to play, along with the BLS, repeat step 3 and see whether the memory is more willing to leave your mind.

TIPS FOR BILATERAL STIMULATION WITH MEMORY REPLAY

As stated previously, don't intentionally bring a trauma to mind in order to practice this technique, as that type of work is best completed with a licensed therapist. However, for times when you cannot seem to let go of thoughts about the trauma, this technique can help you desensitize to the memory and quiet it. Practice this technique when you experience intrusive trauma thoughts for up to five minutes.

Skill 2: Identifying Intrusions (Bottom-Up and Top-Down)

You may notice that specific pieces of intrusive traumatic memories tend to arise more than others. For example, these intrusions may include sounds, images, or smells related to the trauma. In fact, traumatic memories can be reexperienced in any sensory modality. This exercise helps you identify the ways in which you tend to reexperience your trauma. To practice this skill, complete the worksheet to explore the ways in which your trauma intrusions present. You can also visit http://www.newharbinger.com/48848 to download a copy of this exercise.

Trauma Intrusions Worksheet

Consider these questions about how your brain might present trauma reminders to you. As you think about the answers to these questions, you may choose not to write them down due to the distressing nature of the impressions. However, identifying and predicting these reminders can help you better plan for them when they arise.

How do your trauma reminders show up? For some, they may present as images or pictures, or even a movie reel that plays. Write about whether your intrusions tend to include images and whether those images are static or moving.

_____

_____

_____

What sounds do the trauma intrusions include? Some sounds might be of a person's voice, or something a person said, or they may be unrelated to people. You may also recall silence as part of a trauma.

_____

_____

_____

What smells do the trauma intrusions include? This can be one of the more powerful ways a trauma can be reexperienced.

_____

_____

_____

What tactile experiences do the trauma intrusions include? Some examples are heat, cold, a particular texture, pressure, etc. Tactile memories are common but often overlooked in trauma reminders.

_____

_____

_____

What physical sensations do the trauma intrusions include? Do they resemble how you felt at the time of the trauma? For example, racing heart rate, shallow breathing, muscle tension, or lightheadedness may be sensations you felt at the time of the trauma, which you may also feel when reexperiencing the trauma.

_____

_____

_____

What are other ways in which you tend to reexperience your trauma?

_____

_____

_____

When you can quickly recognize your trauma reminders and predict how they may show up, it will likely become easier to catch them, recognize them as being familiar, and manage their intensity.

TIPS FOR IDENTIFYING INTRUSIONS

Identifying intrusions is not a skill you should practice on a regular basis. Rather, it is best to complete this exercise once every several months, to determine how your trauma intrusions might be changing and for the purpose of learning how to become less reactive to them.

Skill 3: Psychological Distancing (Bottom-Up and Top-Down)

Skill 3 builds on the awareness you uncovered in skill 2 and teaches you how to manage these distressing intrusions with psychological distancing techniques, simple strategies that can reduce the impact of, and distress related to, traumatic reexperiencing. This is done by altering the trauma reminders in various ways. Psychological distancing, which is often used in multi-channel eye movement integration (MEMI) therapy (Deninger, 2021), helps you shift your relationship to your trauma reminders so that they become less intense and upsetting. When your reactions to your trauma reminders are less distressing, they can decrease in frequency over time, as your brain will stop signaling them to be something important to pay attention and react to.

Below are several examples of psychological distancing techniques, categorized by domain (images, sounds, sensations, etc.). When you experience traumatic reminders and you have difficulty shifting your attention away from them, these techniques can help neutralize them. Review the Psychological Distancing Techniques below, and then complete the Psychological Distancing worksheet. You can also visit http://www.newharbinger.com/48848 to download a copy of the worksheet.

PSYCHOLOGICAL DISTANCING TECHNIQUES

Psychological distancing for images

- Imagine the image being far away from you, where you're looking at it from a distance.

- Imagine the image is far away from you and shrunk into a small space, like within a picture frame.

- If the image is in color, shift it to black and white.

- If the image is dark or a shadow, imagine it appearing in a bright color.

- If the image is crisp, in clear focus, imagine that it is a picture taken by a moving camera, where it becomes a bit fuzzy.

- If the image is static and unmoving, think of a follow-up image to intentionally bring to mind, to shift away from the original image. This follow-up image can be a neutral image or perhaps an image of the moment the trauma ended and you became safe (if applicable).

- If the trauma images move, as in a movie reel, add on to the end of the "reel" a neutral or positive image symbolizing a sense of safety or relief (if applicable).

- If the image(s) or movie reel is reexperienced as occurring in front of you, where you're looking directly at it, try to reposition it so that it is playing in your left or right periphery. Thus, you are no longer looking directly at it, though you are aware of what is occurring or "playing."

- Look around your environment and intentionally pay attention to what you see around you, as this may dull the traumatic image(s) in your mind.

Psychological Distancing for Sounds

- Imagine the sound(s) being far away, in the distance.

- Imagine the sound(s) is playing along with other sounds, so that you're adding to the trauma sound(s) other neutral or benign sounds in your mind (such as white noise).

- As you notice the trauma sound(s), intentionally bring to mind a neutral or positive image or set of images, as this can create a conflicting experience that dulls your reactivity to the sound(s).

- Imagine the sound(s) while actively listening to other sounds currently in your environment, as this may dull or distract you from the trauma sound(s).

Psychological distancing for smells

- Take deep breaths as you remember the trauma smell(s), as this can send upward signals to your brain reminding you that you are not currently experiencing those smells.

- Imagine the smell(s) while actively smelling something else currently in your environment, as this may dull or distract you from the trauma smells.

- If you begin to actually experience the smells you experienced at the time of the trauma, normalize this experience as a reexperiencing symptom of PTSD, not an indication of psychosis. Additionally, remind yourself that although it seems like the trauma is happening again, there is nothing dangerous or traumatic occurring.

Psychological distancing for tactile experiences

- If your trauma intrusions include feeling tactile experiences you had during a trauma, such as feeling heat, cold, sharpness, another texture, pressure, pain, etc., notice where these tactile experiences are occurring, and intentionally create a new tactile experience by engaging your sense of touch with something in your environment. For example, if your arm begins to feel hot and this is a trauma tactile experience similar to being burned, you may splash cold water on your arm or hold a piece of ice on it. Even simply touching an area that is experiencing a tactile memory can be a helpful reminder to your body that it's no longer experiencing trauma.

Psychological eistancing for physical sensations

- Usually, trauma memories of physical sensations include stress-related reactions in the body, such as rapid heart rate. When these occur, you may practice a relaxation (aka amygdala downregulation) exercise in order to help your body shift out of the stress response that reminds you of the trauma.

Psychological Distancing Worksheet

Answer the following questions to help you assess which type of trauma intrusions you need to focus on.

What type of trauma intrusions do you tend to experience?

Images? What are they? _____

Sounds? What are they? _____

Smells? What are they? _____

Tactile? What are they? _____

Physical sensations? What are they? _____

Which type of intrusion is most distressing or common for you? Which is the least problematic? Below, rank the intrusions from 1 to 5, where 1 = the most distressing type, and 5 = the least distressing type.

_____

When do you tend to experience the most distressing intrusions? Spontaneously? When encountering triggers? Jot down what you've noticed about these intrusions.

_____

_____

_____

Which of the psychological distancing techniques would you like to try to address your trauma intrusions? List below what you'd like to try when encountering the intrusions.

_____

_____

_____

TIPS FOR PSYCHOLOGICAL DISTANCING

You can practice the psychological distancing strategies whenever you experience trauma intrusions that you cannot disengage from. Thus, these are not exercises to focus on whenever you're not bothered by traumatic reexperiencing. Identify the types of intrusions that are most distressing for you (tactile, sounds, etc.) and refer to the specific strategies that address those domains. If you are able to incorporate psychological distancing techniques, they can help you begin to desensitize, and react less to, your trauma reminders.

Skill 4: Distraction Meditation (Top-Down)

When trauma reminders suddenly appear, it can be difficult to stop thinking about them. In addition to the other skills taught in this chapter, one option is to engage in a strong, sensory-based meditation that can help deemphasize the reminders. You can think of this as an extension of psychological distancing, where you learned ways to reduce the intensity of trauma intrusions, except that the goal here is to create and focus on a strong sensory or cognitive distraction that drowns out (to some extent) the intrusion. You can also visit http://www.newharbinger.com/48848 to download a copy of this worksheet.

Distraction Meditation Worksheet

Answer the following questions to help you create a custom meditation for your intrusive reminders.

When you notice you're experiencing an intrusive trauma memory, note whether the memory is mostly comprised of images, sounds, smells, etc. What are the strongest pieces of the intrusion? Are they things you saw, smelled, heard, etc.?

_____

_____

When you've identified the most salient component of the intrusion ("My sister's voice..." or "The sight of the wreck..."), now explore an item, scene, object, or stimulus in your environment, or in your mind, that falls into the same category as the intrusion (image, sound, smell, tactile experience, or physical sensation, etc.). This will become your "distractor." The distractor should produce a strong sensory experience that grabs your attention. Examples of this may include putting your hand in very warm water or applying firm pressure on an area of your body (such as rubbing your temples). If the reminder is a sound, you can identify another sound that you can use to distract from the intrusion, such as music or strong tones. What would you like to try as a distractor?

_____

_____

Now, begin to engage with the distractor by focusing your attention on it (playing the music, feeling the water, looking at the scenery, etc.). Concentrate on the distractor, holding it in mind and doing your best to focus solely on it. When the trauma intrusion enters your awareness, simply notice it, but continue to focus on the distractor. You might imagine the intrusion as something happening in the background, or in your periphery, where you notice it but do not focus on it. When you find your mind has wandered from the distractor, congratulate yourself on this self-awareness and gently redirect your focus back to the distractor. Continue with this meditation for five to ten minutes. What do you notice as you do this?

_____

_____

_____

After completing this meditation, reflect on your experience:

Is the intrusion that presented today one that frequently occurs for you?

_____

_____

Did the distractor help you focus away, and distract from, the intrusion? To what extent?

_____

_____

How did this distractor compare to others you've tried in the past, if any?

_____

_____

When you want to practice this skill again, what distractor would you like to select?

_____

_____

What did you notice about the intrusion as you focused on the distractor? Did the intrusion itself change at all? Did your reaction to the intrusion change?

_____

_____

_____

TIPS FOR DISTRACTION MEDITATION

Practice this skill whenever you experience intrusive trauma memories that are difficult to manage. For some people, five to ten minutes is a sufficient period of time to dull the intrusion and feel less distressed about the intrusive reminder. Also, when selecting your distractor, do not select a sensation or tactile experience that produces pain or is self-harming, as this will create more distress.

Skill 5: Notes to Self (Top-Down)

When trauma reminders take over, you may feel emotionally hijacked. This is because sometimes your amygdala *is* actually hijacking your prefrontal cortex, the area of your brain in charge of logical thinking and reasoning. When this happens, you may feel as though you're drowning in the trauma reminders, unable to think clearly, and it can be difficult to remember the positive perspectives, affirmations, or hopeful thoughts that make you feel better. This exercise has you generate notes, statements, and reminders to reflect on when you feel triggered.

Complete the Notes to Self worksheet when you're feeling grounded, clearheaded, and rational, and when you can access a healthy perspective, so that you can refer back to these healthy perspectives when you experience distressing trauma reminders. Having "notes to self" on hand when intrusions occur can be helpful, especially when used in combination with the other tools in this chapter. You can also visit http://www.newharbinger.com/48848 to download a copy of this exercise.

Notes to Self Worksheet

Answer these questions to help you identify the things you want to bring to mind during trauma intrusions.

When you're in the throes of an intrusive trauma memory, what do you typically remember? Memories may include sensory details or more of a narrative of an event (such as the facts of an event).

_____

_____

_____

_____

How do you tend to feel during the trauma intrusions? Some people feel a sense of hopelessness, terror, horror, or fear.

_____

_____

_____

What do you tend to tell yourself, or believe, during the trauma intrusions? For example, some people believe that they will feel this way forever or that they deserved a traumatic event to happen.

_____

_____

_____

When you reflect on traumatic memories during moments where you are not feeling upset or triggered, how is your perspective on the trauma different? What do you tell yourself in these calmer moments that you can't remember when you're feeling triggered?

_____

_____

_____

What is at least one (or more!) thing that you would like yourself to remember when you experience trauma intrusions? You may, for instance, want to remind yourself that they always pass or that the trauma is over.

_____

_____

_____

If you could talk to yourself during moments of overwhelm using the calm brain you have right now, what would you say to soothe yourself or keep perspective? Write down a narrative of the things you'd like yourself to know, or consider, when you're triggered.

_____

_____

_____

_____

_____

What would your best friend, family member, or other trusted person tell you when you are triggered by trauma reminders? You may have sought comfort from one of these people in the past. If you have, what did they say, if anything, that was helpful to you that you'd like to keep in mind when you feel upset? Or what do you imagine they would say to help you keep a healthy perspective and feel better?

_____

_____

_____

Consider making a video for yourself to watch when you're feeling triggered by trauma memories. In the video, talk directly to yourself, remembering how you feel in those moments when you're overwhelmed by trauma symptoms. What would be helpful to hear from yourself? What advice do you have for yourself in those times? You may remind yourself to practice a relaxation or grounding technique, call a friend, or engage in a self-care activity. You may also share with yourself a "reality check" or some logic that can get lost during traumatic reexperiencing.

TIPS FOR NOTES TO SELF

Be sure to complete this exercise when you're feeling calm and untriggered, as the goal of this activity is to access your rational mind and remember all of the soothing, healthy perspectives that you can. Also, keep some of your most comforting answers accessible, perhaps writing them on notecards or in a notes app on your phone. This way, you can easily access them when you feel triggered by trauma reminders. If these notes are not readily accessible, you're unlikely to remember where to find them when you're feeling overwhelmed.

A Pause to Reflect

Now that you've come to the end of this chapter, you may want to take a moment to pause and reflect on what you learned and which of the skills might be helpful to you as you work on managing your trauma intrusions. The following worksheet can help you organize your thoughts by listing the exercises that you've either found helpful or would like to try in the future. The skills you include can be the ones taught in this book in addition to others outside this book that you may have found useful. You can also visit http://www.newharbinger.com/48848 to download a copy of this exercise.

Skills for Managing Intrusive Trauma Reminders

The following chart is a reminder of the skills in this chapter and a place to record how each skill works for you. When you notice intrusive reminders are negatively impacting your life, choose a skill that is appropriate for the situation. The second column tells you when the best time to practice is. In the third column, make a note of when and how you try the skill over the coming week. Then, take a moment to reflect on your experience. If you have learned other useful skills or techniques that help, add them to the chart in the blank spaces. If you need more space, you can print more pages or keep a journal to reflect on your experiences.

Skill/Technique	Best Time to Practice	When I Practiced the Skill (Day and Time as well as Description of the Situation)	Reflect on the Experience. Was the Skill Useful?
Bilateral Stimulation with Memory Replay	When I am confronted with trauma reminders.		
Identifying Intrusions	When I am experiencing trauma intrusions or outside of those experiences.		
Psychological Distancing	When I am confronted with trauma reminders.		
Distraction Meditation	When I am confronted with trauma reminders.		
Notes to Self	Complete when I am calm and *not* experiencing trauma reminders, to better manage triggers when they occur.		

CHAPTER 2

Feeling Anxious and Revved Up

Angela, a fifty-two-year-old woman, experienced a home invasion almost twenty years ago. While she was not physically harmed during the invasion, the burglar stole several precious items and put her physical safety at risk, leaving her feeling violated and unsafe. Since experiencing the trauma she hasn't been able to let her guard down and relax, even when she logically knows nothing is wrong. Angela has said to her friends and family that she believes she should be "over it" by now, as the burglary happened so long ago, but she can't seem to shake her anxiety. When asked the last time she felt at ease and relaxed, Angela said she can't remember what feeling "normal" is like, as her hypervigilance and anxiety have become a sort of "new normal" now.

Check-In: Your Experiences

After reading about Angela, take a moment to think about your experiences. You may choose to reflect and journal on the following questions:

- Have you ever felt similar to Angela? Perhaps feeling as though you should be "over" something that happened in the past? What's that been like for you?

- Has there been a time when something terrible happened and after that you had a hard time relaxing? Or maybe you noticed feeling more anxious after that event, even when you're not in danger? How has that impacted you?

- Do you ever worry that feeling anxious and revved up will be your "new normal" or that your ability to relax has been stolen? How have those concerns changed you or how you live your life?

How Feeling Anxious May Impact Your Life

Feeling chronically anxious, on guard, and hypervigilant can impact your life in many ways, and may impact multiple domains such as work, relationships, and hobbies. Sometimes it can be difficult for trauma survivors to recognize their own anxiety and hypervigilance, as feeling revved up can easily become a way of life or a sort of emotional habit that occurs outside of conscious awareness. This was the case for Angela.

However, there are several indicators of hyperarousal, which can be thought of as a type of anxiety, that trauma survivors should be aware of. Some of the most common indicators of hyperarousal and anxiety include an inability to relax, focus, sleep well, or feel at ease in the body. You might feel nervous or fearful or notice changes in your thinking, where you worry about bad things that might happen, imagine worst-case scenarios, or repeat the same thoughts over and over (rumination). These thoughts, or cognitions, can interact with, or even cause, the physical sensations and emotions associated with anxiety.

The following brief self-assessment, which is adapted from the Beck Anxiety Inventory and the *DSM-5* (American Psychiatric Association 2013; Beck et al. 1988), is a fast way to determine whether you might be experiencing hyperarousal and anxiety symptoms. (Note: The *DSM-5* is the "therapist bible" that defines the criteria for all psychological diagnoses.) You can also visit http://www.newharbinger.com/48848 to download a copy of this exercise.

Hyperarousal and Anxiety Self-Assessment

Read the following and circle the best response from 0 to 3, where 0 = None/not at all, 1 = A little bit/ sometimes, 2 = A moderate amount/often, and 3 = A lot/most of the time.

I have a hard time falling or staying asleep at night.	0	1	2	3
I can't relax.	0	1	2	3
I'm afraid of losing control.	0	1	2	3
It is difficult for me to concentrate on tasks.	0	1	2	3
I have to reread written material, as I do not "digest" it well.	0	1	2	3
I feel on guard even when it doesn't make sense to be vigilant or when it's unhelpful to me.	0	1	2	3
I worry a lot or imagine bad things that might happen.	0	1	2	3
I snap at people without meaning to.	0	1	2	3
I lose control of my temper.	0	1	2	3
I find myself scanning my environment, looking for something that might be wrong.	0	1	2	3
I am jumpy and startle easily.	0	1	2	3
My mind spins when I try to go to sleep.	0	1	2	3
I feel a buzzing in my body.	0	1	2	3
My body feels tense, hot, and/or shaky.	0	1	2	3
My breathing is fast and shallow.	0	1	2	3
My heart feels like it is pounding.	0	1	2	3
I replay things over and over in my mind and sometimes overanalyze them.	0	1	2	3
I feel nervous, unsettled, and/or uneasy.	0	1	2	3
I feel dizzy or light-headed.	0	1	2	3
My chest or throat feels tight.	0	1	2	3
I feel free-floating anxiety, where I feel like something bad might happen, but I'm not sure what.	0	1	2	3
I feel irritable and on edge even when nothing is wrong.	0	1	2	3
I take risks that may be harmful to me, such as driving too fast, spending more money than I have, or intentionally hurting myself.	0	1	2	3
Total				

Add up the total score from the above items, and record it next to "Total." A score of 24 or higher indicates elevated hyperarousal/anxiety symptoms.

While some of these symptoms can be subtle, they can affect your ability to function optimally, as they can make it difficult to focus, sleep, perform well, and enjoy healthy, happy relationships. Use the following exercise to reflect on your self-assessment score. You can also visit http://www.newharbinger.com/48848 to download a copy of this exercise.

Reflect on Your Self-Assessment Score

What did you score on the self-assessment? Was this surprising to you? Take a moment to reflect on, and process, what you just learned about yourself. You may consider the following questions as you reflect on your experience with this.

Was your score higher than you thought it would be? Lower?

_____

_____

What does your score indicate about you and the challenges you're experiencing?

_____

_____

_____

Were there items on the assessment you hadn't considered before?

_____

_____

_____

Were there items that surprised you? Were there things you hadn't been associating with hyperarousal/anxiety?

_____

_____

_____

How do you think these symptoms have been impacting your life?

_____

_____

_____

Why Hyperarousal After Trauma Is So Common

Traumatic experiences frequently result in "one trial learning," meaning that a single event, or experience, can alter brain structure and/or functioning, which in turn results in new learning that is quickly solidified. Though other types of learning usually require repetition, trauma learning can happen after just one experience, just like in Angela's case. This new learning, while usually very helpful for survival, can compromise mental health and long-term well-being.

In particular, after trauma the brain may learn that it is unsafe to relax, that the world is (or people are) dangerous, and/or that constant vigilance is protective. For instance, if you experienced a home-invasion incident like the one Angela experienced, your brain may "learn" that you are not safe in your own home, or that you are unsafe when you are alone, or that you need to constantly be on guard for intruders. The ability to learn these lessons, and quickly, demonstrates how powerful the brain can be!

During traumatic experiences, the "threat detection" center of the brain—the amygdala—activates, and this leads to a cascade of reactions that put the brain and body in the stress response. This "survival mode," when activated, leads you to (unconsciously) select an instinctive survival strategy, which may include fighting, fleeing, or freezing. And these survival responses, all of which are associated with amygdala activation and stress, become credited with your survival—whether they actually helped or not. In other words, your brain learns that you survived the trauma, essentially, by *becoming stressed* and *entering survival mode.*

Given that the brain's primary goal is to keep you alive (not mentally healthy!), it makes sense that the brain learns, very quickly, how to best do this when your life is threatened. Moving forward, after a traumatic event, the brain has learned that threats to your safety exist, and that the best way to keep you alive in the face of those threats is to keep you hyper-attuned to your surroundings at all times—even if that means you're stressed out, anxious, and in survival mode all the time. The result is that you may find it difficult to exit survival mode, downregulate the amygdala, and just relax.

To better understand how anxiety/hyperarousal occurs after trauma, here is a summary of what many people experience:

1. **Trauma happens, and the amygdala activates.** When you experience trauma, your amygdala, which is the area of your brain responsible for detecting threat and danger, activates. This activation alerts the rest of the brain and the body that a threat is present and something terrible is happening (or about to happen). This is what happened when Angela first realized someone was breaking into her home.

2. **Your stress pathway activates.** After the amygdala activates, it sends the threat signals to the hypothalamus, which then activates the "stress pathway" known as the HPA axis. The HPA axis is comprised of the hypothalamus, pituitary, and adrenals. When this pathway becomes activated, it initiates the stress response in the body.

3. **Your body goes into the stress response.** When the HPA axis activates, the body enters the stress response, which often leads to somatic reactions such as a racing heartbeat, shallow breathing, muscle tension, upset stomach, dizziness, and other symptoms. Angela may have experienced these, and other symptoms, as the trauma occurred.

4. **The "thinking area" of the brain dulls.** As the body moves into the stress response, the amygdala's activation also dulls activation in the "thinking area" of the brain, the prefrontal cortex. The result is that it becomes difficult to focus, concentrate, or

think clearly. It also can lead to difficulty remembering some pieces of the traumatic event.

5. **The brain chooses a survival response.** In an attempt to survive the threat the amygdala has detected, the brain will choose a survival response—fight, and prepare to address the threat head-on in some way; flee, and run away or escape the threat somehow; or freeze, and go numb and immobile in the face of something that seems or is too overwhelming to respond to otherwise. Angela reported that during the traumatic event she fled, running out the back door in an attempt to escape danger. Fleeing was the stress/survival response her brain chose.

6. **You survive the trauma, and the brain remembers how.** When the traumatic event is over and it becomes clear that you have survived, the brain makes note of this and quickly learns that the survival response (which occurred in the context of amygdala activation and extreme stress) worked. Fighting, fleeing, or freezing, and becoming extremely stressed, helped keep you alive. To ensure survival moving forward, the brain continues to produce the stress response, making you feel chronically anxious and hyperaroused. While this feels terrible, the brain has learned that the world can be a dangerous place and that survival responses can keep you alive, so the amygdala continues to activate, keeping you in "survival mode" just in case another threat comes along. Angela faced this firsthand, as she experienced chronic anxiety, hypervigilance, and fear after the home invasion.

Use the following exercise to reflect on your survival responses. You can also visit http://www.newharbinger.com/48848 to download a copy of this exercise.

Your Survival Responses

We each have our own go-to survival responses when we get stressed and anxious. Which responses do you notice in yourself, and when? Answer the following questions to better understand your own survival responses.

The Fight Response: Those with a strong fight response face anxiety and overwhelm head-on, becoming hyperproductive or insisting that issues with other people be dealt with immediately. They may also be confrontational or assertive when anxious or angry. Do you ever enter the fight response? When?

Times when I tend to fight: _____

_____

The Flight Response: It's common for some people to disengage when stressed or anxious, which may be a sort of flee, or flight, response. This may be apparent when a person physically leaves a situation, distracts themselves by becoming immersed in their phone, or ignores problems they have. Do you ever flee? When?

Times when I tend to flee: _____

_____

The Freeze Response: Some people shut down when they're overwhelmed and become unproductive, quiet, or nonresponsive, which can be thought of as a sort of freeze response. Sometimes this also happens when a person forgets what they want to say when stressed out, such as during an argument. Do you ever freeze? When?

Times when I tend to freeze: _____

_____

You may notice from your responses that you're prone to experiencing one particular survival response, which you can think of as your "go-to response" when anxious or overwhelmed. Knowing this about yourself can be helpful, as you can communicate these tendencies to others and understand why you may act the way you do when stressed.

Terms Your Therapist Might Use to Explain This Challenge

If your experiences are similar to Angela's, your therapist may use the following descriptors and terms in your discussions, or in your therapy notes:

- **Anxiety.** This is a general descriptor that can refer to a particular disorder, such as panic disorder, or a category of disorders (anxiety disorders). It can also refer to a feeling state ("I'm feeling anxious"), where there is no formal diagnosis present. Symptoms of anxiety can include feeling worried, revved up, on guard, nervous, and/or shaky. Additionally, arousal and reactivity symptoms, described next, are often conceptualized as anxiety symptoms.

- **Arousal and reactivity symptoms.** This is one of the symptom clusters of PTSD according to the *DSM-5*, and it is comprised of six symptoms, including irritability/aggression, destructive behavior, hypervigilance, sensitive startle response, difficulty concentrating, and difficulty sleeping. Although trauma survivors often do not experience all of these symptoms, it is common for individuals to report one or more of them.

What Your Brain Needs

The good news is that several areas of the brain involved in mental health and illness can be altered (structurally and/or functionally) in positive ways, through psychotherapeutic techniques, medication, and the consistent practice of straightforward tools and techniques. Just as your brain learned how to be on guard, anxious, and hypervigilant, it can learn how to become calm. If you feel nervous and on edge and have difficulty with sleep, attention, and racing thoughts, here is what your brain needs to feel more relaxed.

1. Bottom-up techniques that reduce the activation of the threat detection center (amygdala). As described earlier in this chapter, when the amygdala activates, it sets off a chain of reactions that produce the stress response in the body and the brain. If your goal is to feel less stressed and less anxious, then you need to reduce activation of the amygdala, which will in turn allow the body and brain to relax. Thankfully, several bottom-up techniques, where you work through the body to change the brain, can help lower amygdala activity and reactivity.

2. Top-down techniques that mobilize attention and deemphasize focus on anxious thoughts. While bottom-up techniques are the best and fastest way to downregulate the amygdala, top-down techniques, where you work with your thoughts to change the brain (instead of working with the body) are also helpful in teaching your brain to relax. When you practice a technique that has strong top-down components, you learn how to strengthen the thinking, rational areas of your brain. In turn, strengthening these brain regions can help you downregulate the amygdala with your mind.

The next section details five skills that contain both bottom-up and top-down components that can help you learn how to downregulate your brain's threat detection center. When you are able to reduce your amygdala's activation, it will allow you to feel less anxious and vigilant, and calmer and more relaxed.

Skills for Meeting the Challenge

In this section, you'll learn five skills that can help you learn to ground yourself and reregulate when you enter a state of hyperarousal. These skills include deep breathing, breathing in ways that activate the mind-body connection, the body-based relaxation technique known as autogenics, activating the vagus nerve using the diving response to reduce amygdala activation, and activating mirror neurons in the brain to help you connect to others and be soothed by them.

Skill 1: Deep Breathing (Bottom-Up)

This exercise is categorized as a bottom-up technique, where you will calm the amygdala by working with the body, or in this case, the breath specifically. While it has become a cliché to simply say, "Take a deep breath," or "Just breathe!" the breath remains a very powerful way to engage the vagus nerve and reduce amygdala activation and stress. However, some people are hesitant to commit to breathing exercises and may not understand how they can be helpful. In particular, there are a few reasons that breathwork is often met with resistance or is believed to "not work."

The first is that many people do not breathe in a way that actually promotes relaxation. Rather, not realizing it, they breathe in a way that actually worsens their anxiety. When you breathe, your mind may spin on scary or otherwise distressing thoughts, and this counteracts the relaxing effects of the breathing exercises. Even if you're breathing correctly for relaxation, if your mind is telling you that you're going to die in a plane crash, it will be very difficult to calm down. Finally, breathing may not work as well as desired if you wait until you're almost panicking to practice it. While deep breathing can be helpful in moments of acute stress if you are very skilled at it (meaning, you've been practicing it consistently for at least a few months), it is more helpful at reducing longer-term, chronic stress and anxiety than severe, acute anxiety. Therefore, practice deep breathing as a daily skill, rather than using it only during moments of strong anxiety. Here are some instructions that will teach you how to breathe correctly for stress management and help you get the most out of your breathing practice.

1. **Strike a pose.** A key reason that breathing may not help you relax is that you may be practicing breathwork while in a pose, or stance, that makes it difficult to activate the vagus nerve and relax your body. In particular, many people attempt deep breathing while slouched over or with their shoulders curved inward. These defensive stances can lead to shallow, restricted breathing that actually worsens anxiety. Thus, before you even begin your deep-breathing exercise, make a note of your body's position and consider assuming a pose that will maximize your chances of being able to breathe fully and deeply, activating the vagus nerve. Here are some examples of poses that will increase your chances of feeling relaxed when you breathe. Choose one and make it a habit to strike this pose when you engage in deep breathing!

   Pose 1: Sit up with your back straight and gently put your hands behind your head, feeling your rib cage being pulled out.

   Pose 2: Sit up with your back straight and place your hands on your rib cage, keeping your shoulders down and back.

Pose 3: Lying down flat on your back, rest your arms at each side, palms up.

2. **Begin to breathe.** Next, begin taking slow, long, deep breaths in and out, just noticing the quality of the breath and any sensations. Focus on inhaling and exhaling deeply, but only inhale as deeply as is comfortable to you. When you exhale, try to elongate it, extending it longer than the inhale. Extending your exhales is an easy way to relax faster while breathing.

3. **Focus the mind.** In order to distance yourself from distressing or ruminative thoughts, begin to focus the mind, intentionally, on a word, thought, or phrase. Or many people choose to focus the mind on counting and count up to a certain number on each inhale and exhale. For example, you may count to five as you inhale and count to eight as you exhale (to elongate and emphasize the exhale). Take several breaths like this, maintaining the pose, focusing the mind, and extending the exhales.

4. **Incorporate breath holding.** Next, you may begin to incorporate some gentle breath holding. This need not be uncomfortable. Simply begin to briefly hold your breath after each inhale, for a count of two to four (whatever is comfortable to you). Repeat this on each inhale before initiating the exhale. Holding the breath after each inhale sustains oxygen in the brain and keeps pressure on the vagus nerve, which allows the breathing exercise to relax you more fully, and faster.

5. **Repeat.** Now, repeat the breaths for five to ten minutes, focusing on full, deep inhales, holding the breath for a few seconds, and exhaling slowly, all while staying in the pose you selected. Here's an example of how this might be practiced:

Position your hands behind your head, and now begin to inhale slowly for a count of 1 … 2 … 3 … 4 …

Now hold the breath for a count of 1 … 2 … 3 … 4 …

And now begin to very slowly exhale for a count of 1 … 2 … 3 … 4 … 5 … 6 …

Now repeat…

TIPS FOR DEEP BREATHING

Practice, practice, practice! Practice makes progress. If you adopt a regular breathing practice (daily is ideal), over time you will notice that the breathing works better, and faster, as you become skilled at it. Your amygdala will learn, with repeated practice, how to downregulate quickly as a result of the breathing, meaning that this skill will likely become increasingly more powerful and helpful over time. Aim to focus on deep breathing for five to ten minutes each day as a formal practice. Additionally, you can informally practice deep breathing when engaging in other tasks such as driving, working on a computer, or watching television. The more you train your body and brain to breathe fully and deeply, the more it will become a habit.

Skill 2: Mind-Body Connection Breathing (Bottom-Up and Top-Down)

Much like the last skill, this exercise is a breathing technique, which is categorized as a bottom-up technique. However, this technique is more complex, incorporating top-down components such as imagery, which makes it a powerful bottom-up/top-down combination exercise. Specifically, this technique integrates deep breathing, vagus nerve activation, attention to the breath, visual imagery, thought management, and body positioning to help you achieve a state of relaxation and amygdala deactivation. To practice this exercise, follow these instructions.

1. **Assume a relaxing position.** Begin by positioning your body according to one of these poses:

    Pose 1: Lying flat on your back, prop your lower legs up on a chair or some other surface that allows your knees to bend at a 90-degree angle. Your arms may rest at your sides or you may place them above your head.

    Pose 2: Lying flat on your back, bend your knees and place your feet flat on the floor. Your arms may rest at your sides or you may place them above your head.

    Pose 3: Locate a blank wall, lie on your back with your rear facing the wall, and gently lift your legs and scoot toward the wall until your legs are straight up the wall, at a 90-degree angle.

Each of these poses accomplishes an important objective: psoas relaxation. Your psoas muscles connect your spine to your legs and your pelvis to your torso. Importantly, these muscles are believed to react to, and store, stressful and traumatic experiences, as they tense when you're feeling stressed out (and during moments of trauma). Thus, one way to work with the body to relax the amygdala is to release psoas tension, which communicates to the brain that no threat is present and it is okay to relax.

2. **Bring your attention to the breath.** Now in your relaxed pose, gently close your eyes and bring your awareness inward, focusing on the breath. At this point you are not changing the breath; rather, simply notice what it feels like to inhale and exhale. Stay here for a moment, just being with the breath.

3. **Visualize the breath.** As you stay connected with the breath, begin to incorporate visual imagery by imagining that the air you inhale is one color and the air you exhale is a different color. These can be any colors you choose—for instance, you may choose to inhale purple air and exhale orange air—as long as they're colors you like and that you can associate with relaxation.

4. **Incorporate a mantra.** Next, as you visualize the breath in colors, you may slowly repeat a word, mantra, or phrase as you inhale and as you exhale. You may use the same word for each or select different words for the inhale and exhale. For example, you may inhale "energy" and exhale "calm." Focusing the mind on a particular word is an effective way to distance from distressing and ruminative thoughts that might be revving up your stress.

5. **Extend the breath.** Continue visualizing the breath and repeating the word(s)/mantra(s) as you remain in the psoas relaxation position, but also begin to extend the breath, taking longer, fuller inhales and exhales. This, along with your body position, will help activate your vagus nerve, which is a cranial nerve that promotes relaxation in the body and brain. Continue with this exercise for about ten minutes.

TIPS FOR MIND-BODY CONNECTION BREATHING

This exercise can be modified as desired if you find that it is too challenging or difficult to integrate all the components. For instance, you may alter the technique so that you complete it sitting up, if lying down is triggering for you. Or if you experience difficulty with imagery or experience intrusive images when you close your eyes, you may opt to complete this exercise with your eyes open. Document or journal the specific variant of the technique that works best for you, and feel free to be creative as you discover what is most helpful. The most important parts of this technique, which should not be altered, include deep breathing, vagus nerve activation, and identifying something to focus the mind on (whether that is a calming thought, word, image, etc.).

Skill 3: Autogenic Training (Bottom-Up and Top-Down)

Autogenic training is a combination bottom-up and top-down technique that contains both cognitive and somatic elements. In this exercise, you will focus on different regions of the body and, while attending to these areas, repeat statements about how these areas feel.

You may or may not actually *feel* the statements you say to yourself, and that is okay. The main purpose of this technique is to learn how, with your mind, to redirect blood flow in a manner that is consistent with the relaxation response, instead of the stress response. When the amygdala activates and the stress pathway (HPA axis) turns on, it produces the stress response in the body. This stress response is associated with changes in blood flow, where blood rushes away from the abdominal area, hands, lower arms, feet, and lower legs, and rushes deep into large muscles and the head, neck, shoulders, and back. This change in blood flow makes you stronger and faster, so that you can fight or flee life-threatening dangers. However, when there is no danger, this blood flow redirection is not helpful and is associated with gastrointestinal issues, chronic muscle tension and pain, and headaches. In this exercise, you will learn how to redirect blood flow away from your head and out of your large muscles and back to your core region, arms, hands, legs, and feet. Blood flow in these areas is associated with deep relaxation. To practice autogenic training, follow these steps.

1. With your eyes open or closed, bring your attention to the breath, just noticing what it feels like to breathe in and out.

2. Take a moment to scan your body, noticing any tension or discomfort in any area. Just notice this without judgment.

3. Next, shift your attention down to your feet, feeling into this area and noticing any sensations in this region. Also note whether your feet feel cold or warm.

4. Now, state to yourself, either silently or in a soft voice:

    My feet are warm.

    My feet are warm.

    My feet are warm.

    My feet are warm.

    My feet are warm.

    And, may I feel relaxed.

5. Continuing to hold your focus on your feet, now repeat silently or in a soft voice:

    My feet are heavy.

    My feet are heavy.

    My feet are heavy.

    My feet are heavy.

    My feet are heavy.

    And, may I feel relaxed.

6. Begin to gently shift your attention up a bit, noticing any sensations in your lower legs. Focusing on your legs, repeat:

    My legs are warm.

    My legs are warm.

    My legs are warm.

    My legs are warm.

    My legs are warm.

    And, may I feel relaxed.

7. Continuing to hold your focus on your legs, now repeat:

   My legs are heavy.

   My legs are heavy.

   My legs are heavy.

   My legs are heavy.

   My legs are heavy.

   And, may I feel relaxed.

8. Now shift your awareness upward to the hands, beginning to feel into this region, noticing whether your hands feel cold or warm. Noticing your hands, repeat to yourself:

   My hands are warm.

   My hands are warm.

   My hands are warm.

   My hands are warm.

   My hands are warm.

   And, may I feel relaxed.

9. Continuing to focus on your hands, repeat:

   My hands are heavy.

   My hands are heavy.

   My hands are heavy.

   My hands are heavy.

   My hands are heavy.

   And, may I feel relaxed.

10. Next, moving your attention up to your lower arms, repeat to yourself:

My arms are warm.

My arms are warm.

My arms are warm.

My arms are warm.

My arms are warm.

And, may I feel relaxed.

11. Still focusing on your arms, repeat:

My arms are heavy.

My arms are heavy.

My arms are heavy.

My arms are heavy.

My arms are heavy.

And, may I feel relaxed.

12. Next, move your attention to your core, to your abdominal area, feeling into this region and noticing any sensations or experiences. Now, repeat the following:

My stomach is warm.

My stomach is warm.

My stomach is warm.

My stomach is warm.

My stomach is warm.

And, may I feel relaxed.

13. Continuing to feel into the abdominal area, now repeat:

My stomach is soft and relaxed.

My stomach is soft and relaxed.

My stomach is soft and relaxed.

My stomach is soft and relaxed.

My stomach is soft and relaxed.

And, may I feel relaxed.

14. Finally, shift your focus up to the top of your body, to your head and face. Feeling into this area, notice any sensations and repeat to yourself:

My head is cool.

My head is cool.

My head is cool.

My head is cool.

My head is cool.

And, may I feel relaxed.

15. Still noticing any sensations in your head region, now repeat:

My head and face are relaxed.

My head and face are relaxed.

My head and face are relaxed.

My head and face are relaxed.

My head and face are relaxed.

And, may I feel relaxed.

16. To end this exercise, shift the focus to the breath and take two deep breaths before coming back into the room.

TIPS FOR AUTOGENIC TRAINING

In this technique, the order of body regions you focus on is not so important, but the language you use *is*. Do not change the language of the script (changing warm to cool, for instance), as the words used are intended to facilitate blood flow changes that specifically promote relaxation. Changing the language may render the technique less useful for relaxation.

Do this exercise in a quiet space, with minimal distractions, as a formal practice that best allows you to fully focus on it. While you may complete it using the exact script provided, you may repeat each of the statements more than five times if you wish to extend the practice, making it longer and moving even deeper into relaxation.

Skill 4: The Diving Response (Bottom-Up)

This rather counterintuitive, bottom-up technique provides an extremely fast way to calm the body and brain through activating the vagus nerve (described earlier in this chapter). If you are experiencing acute, high-level stress or anxiety, this technique may be especially helpful. Called the "mammalian dive reflex," or simply the "diving response," this technique, though slightly uncomfortable, reduces the stress response and induces the relaxation response. Here's how you do it:

1. **Get cold water or ice.** You may fill a bowl with ice water, start running cold water in the sink, or grab some ice packs (four is ideal). You may also put ice cubes in plastic baggies.

2. **Hold your breath.** This doesn't need to be for long; begin holding your breath and then immediately move to step 3.

3. **Put the cold water/ice on your face.** If you are running water or have water in a bowl, gently begin to splash the water on your face. If you are using ice (which is less messy!), pick up the ice packs and press them firmly on your face, placing one on the forehead, one on your chin, and one on each cheek, holding them in place. Hold

the ice on your face, or continue splashing the cold water on your face, for around thirty seconds.

4. **Release your breath.** Now, begin to breathe again as you stop splashing your face or remove the ice packs. Repeat as desired.

Although this exercise may seem odd, it is a primitive, effective way to induce parasympathetic arousal (which is the technical name for the relaxation response) through vagus nerve activation. It is believed that the body responds to sudden, cold water by lowering blood pressure and heart rate (which make you feel relaxed!) in an attempt to help you survive falling into cold, deep water without warning. The lowered blood pressure and heart rate help you preserve energy you might need if you suddenly find yourself immersed in water with the threat of drowning. This survival reflex can thus be used as a sort of anxiety "hack" when you find yourself revved up.

TIPS FOR THE DIVING RESPONSE

This quick and easy technique may be especially helpful during moments of intense stress or anxiety, such as before a high-stakes meeting, interview, or competition. Some people report that the technique not only calms them down, but also facilitates clarity of thought and focus, which is not surprising given that concentration is much easier when you are relaxed. Additionally, this exercise may be useful to do before bed, as a way to calm down and relax before going to sleep.

Skill 5: Activate Mirror Neurons (Bottom-Up)

A natural, simple, and underused way of reducing anxiety and amygdala activation is to activate the mirror neurons in your brain. Mirror neurons are special cells (neurons) in your brain that allow you to connect with, and attune to, other people. When you activate mirror neurons, you're likely to feel a stronger attachment to the other person. And, as long as that connection is positive and supportive, the mirror neuron activation will also boost oxytocin, sometimes called the "cuddle hormone." When oxytocin increases in your body, it downregulates, or reduces cortisol, which is a stress hormone that is released when your stress pathway (HPA axis) activates. Thus, mirror neuron activation, which increases oxytocin, is a great way to reverse the stress response through lowering cortisol in your system, which in turn helps produce relaxation. Thankfully, there are many ways to activate mirror neurons and boost oxytocin. Here are some recommendations:

- **Connect with other people.** Mirror neurons are activated when we are around others in three ways: through physical touch, sound (such as hearing the voice of another person), and eye contact. To activate mirror neurons and reduce cortisol, make it a priority each day to spend time with supportive people who you can make eye contact with as you talk. Making time for physical contact, such as hugs, with loved ones is also a great way to reduce cortisol quickly. Aim for at least one ten-minute conversation with a supportive family member or friend each day, in addition to frequent physical contact.

- **Connect with animals.** Pets, and animals more broadly, can also activate mirror neurons and produce a sense of calm. If you have pets at home, be sure to touch, pat, and spend time with them daily. Not only will they enjoy receiving the love, but you may also notice feeling less stressed when you are around them!

- **People watch.** While it's preferable to have direct interaction with others for many reasons, another way to activate mirror neurons is to simply *be around* and watch other people. For instance, you may people watch at an outdoor café or park. Though you may not get your social needs met this way, merely watching what other people are doing has been shown to increase mirror neuron activation and produce a sense of calm.

- **Watch a "feel good" movie.** Watching other people on a screen will not produce the same effects as being with others in person. However, you can activate mirror neurons when you watch movies and connect to the characters, and this can affect your mood. Just like watching a scary movie may make you feel scared, watching a "feel good" movie can activate mirror neurons and make you feel warm and fuzzy. This in turn can reduce cortisol and your overall stress level.

TIPS FOR ACTIVATING MIRROR NEURONS

Keep in mind that interactions with others can activate mirror neurons in both positive and negative ways. For instance, if you find that you are in a toxic relationship, the mirror neuron activation you experience with that person may lead you to feeling sad, depressed, or angry. In other words, not all mirror neuron activation is positive! For the activation to work for you, and not against you, and for it to reduce stress and amygdala activation, your interactions with others need to be positive. Finding and interacting with supportive others is critical to your mental well-being!

A Pause to Reflect

Now that you've come to the end of this chapter, it is a good time to pause and reflect for a bit on what you've learned and the skills that you've found helpful. Note what works for you and what doesn't. When we're feeling anxious, the amygdala suppresses activation in the thinking areas of our brain; in those moments, it can be difficult to think clearly and remember the things that are good for us! While we know most of the time that deep breathing may be helpful, for instance, this can be hard to remember while in the throes of anxiety. What we "know" can be difficult to access when stressed.

Thus, it is helpful to write down your coping skills and the things you want to remember to do, so that when you are stressed and not thinking clearly, you have something to refer to. One way to do this is to use the following worksheet to note the skills that help you feel less anxious and more relaxed. This can include skills taught in this book as well as ones you may have learned from other sources. You can also visit http://www.newharbinger.com/48848 to download a copy of this worksheet.

Skills for Managing Stress and Anxiety

The following chart is a reminder of the skills in this chapter and a place to record how each skill works for you. When you notice stress and anxiety are negatively impacting your life, choose a skill that is appropriate for the situation. The second column tells you when the best time to practice is. In the third column, make a note of when and how you try the skill over the coming week. Then, take a moment to reflect on your experience. If you have learned other useful skills or techniques that help, add them to the chart in the blank spaces. If you need more space, you can print more pages or keep a journal to reflect on your experiences.

Skill/Technique	When to Practice	When I Practiced the Skill (Day and Time as well as Description of the Situation)	Reflect on the Experience. Was the Skill Useful?
Deep Breathing	When feeling upset or anxious, but also when feeling calm, as a regular practice.		
Mind-Body Connection Breathing	When feeling upset or anxious, but also when feeling calm, as a regular practice.		
Autogenic Training	When feeling upset or anxious, but also when feeling calm, as a regular practice.		
Diving Response	When feeling upset or anxious.		
Activating Mirror Neurons	Practice on a regular basis, with supportive, positive people.		

CHAPTER 3

Difficulties with Anger

John, a thirty-seven-year-old architect, recalls how growing up with an alcoholic father left him feeling helpless and scared. When his father would bully his mother and siblings, sometimes even hitting them, he swore he would never be anything like that. However, now as an adult, he fears he is becoming to his children what his own father was to him: loud and scary. John says that when he gets stressed with work, or feels like his home is chaotic with messes and noise, he begins to feel out of control and ends up yelling at his family. He says it's gotten to the point where he can't trust himself, as even small things trigger his anger and set him off. This, in turn, has caused him to withdraw from his family on the weekends. While he knows that his trauma history likely plays a role in his current behavior, he feels helpless to get it under control.

Check-In: Your Experiences

After reading about John, take a moment to think about your experiences. You may choose to reflect and journal on the following questions:

- Have you ever felt similar to John? Perhaps feeling out of control, or "triggered," to where you experience difficulty managing strong emotions? What's that been like for you?

- Since experiencing trauma, do you at times feel like you're not who you'd like to be? Do you find yourself feeling or doing things that seem "out of character"?

- Do you find it hard to trust yourself because you're unsure of how you might react in different situations? Have you isolated or avoided people so that you won't blow up, become irritated, or act impatiently toward them?

How Anger Difficulties May Impact Your Life

Many people experience difficulties with controlling anger after trauma. In fact, this is one of the more common complaints survivors report, though it's not always the case that they link these difficulties to the traumatic event(s). Instead, some people attribute their anger to their personality, believing it's "just the way they are." While it's true that each of us has a temperament, dysregulated anger, which manifests in problematic behaviors, is something that can be improved. When anger difficulties go unaddressed, they can cause problems in your relationships and work life. Because of their impact on functioning, anger issues are a common reason that clients seek psychotherapy.

It's often the case that those who experience difficulty controlling their anger know it. However, sometimes it can be difficult to disentangle what is attributable to *you* rather than your environment, your context, or the people around you. Although your behavior is always your responsibility, sometimes trauma survivors believe that their anger is unjustified or that they are expressing it in a harmful way when in fact they are not. In other words, trauma survivors often believe their anger to be problematic or unacceptable, but this is not always accurate. Anger is not always unhealthy!

The following brief self-assessment, the Dysregulated Anger Self-Assessment, is adapted from the Clinical Anger Scale (Snell et al. 1995) and State-Trait Anger Expression Inventory (Spielberger et al. 1983) and is intended to help you determine whether your anger has become harmful or problematic for you or others. This self-assessment details several observations trauma survivors report about themselves. You can also visit http://www.newharbinger.com/48848 to download a copy of this assessment.

Dysregulated Anger Self-Assessment

Read the following and circle the best response from 0 to 3, where 0 = None/never, 1 = A little bit/sometimes, 2 = A moderate amount/often, and 3 = A lot/most of the time.

I feel angry in general (regardless of whether there is any expression of it).	0	1	2	3
I am angry that I have failed in life.	0	1	2	3
I feel hostile toward others.	0	1	2	3
People are trying to annoy me.	0	1	2	3

I am angry that others have messed up my life.	0	1	2	3
I get so angry that I would like to hurt others (regardless of whether there is any expression of it).	0	1	2	3
I get so angry that I *do* hurt other people, emotionally.	0	1	2	3
I get so angry that I *do* hurt other people, physically.	0	1	2	3
I am worried that I will lose control and hurt someone when angry (either emotionally or physically).	0	1	2	3
I yell at people more than I used to, or more than I should.	0	1	2	3
I isolate away from others because of my anger.	0	1	2	3
I have gotten in trouble at work because of my anger.	0	1	2	3
I regret the way I behave when I become very angry.	0	1	2	3
I am impatient.	0	1	2	3
I express my anger by slamming doors, throwing things, yelling, or otherwise striking out.	0	1	2	3
I have been banned or "kicked out" of businesses or other people's homes due to my anger outbursts.	0	1	2	3
I feel out of control when I become angry.	0	1	2	3
I have lost relationships, or almost lost relationships, as a result of my anger outbursts.	0	1	2	3
I know that my anger outbursts are unreasonable or disproportionate.	0	1	2	3
My anger disrupts my ability to make good decisions.	0	1	2	3
I "snap" with no warning.	0	1	2	3
Because of my anger outbursts, other people don't like me.	0	1	2	3
My anger keeps me awake when I'm trying to sleep.	0	1	2	3
My anger is hurting my health (contributing to stomachaches, headaches, etc.).	0	1	2	3
I am surprised by the intensity of my anger.	0	1	2	3
My anger compromises my ability to think clearly.	0	1	2	3
People around me have told me they think my anger is a problem.	0	1	2	3
Total				

Add up the total score from the above items, and record it next to "Total." A score of 28 or higher indicates you may be experiencing problematic clinical anger, and your expressions of anger may affect your ability to enjoy healthy relationships and perform well at work.

Use the following exercise to reflect on your self-assessment score. You can also visit http://www.newharbinger.com/48848 to download a copy of this exercise.

Reflect on Your Self-Assessment Score

What did you score on the self-assessment? Was this surprising to you? Take a moment to reflect on, and process, what you just learned about yourself. You may consider the following questions as you reflect on your experience with this.

Was your score higher than you thought it would be? Lower?

_____

_____

What does your score indicate about your anger and your ability to manage it?

_____

_____

_____

Were there items on the assessment you hadn't considered before?

_____

_____

_____

Were there items that surprised you? Were there things you hadn't been associating with dysregulated anger?

_____

_____

_____

How do you think these symptoms or experiences have been impacting your life?

_____

_____

_____

Why Anger After Trauma Is So Common

During a traumatic event, the brain prioritizes survival by activating the stress pathway (located
in both your body and brain) and very quickly selecting a survival response. In particular, low
subcortical brain regions—especially the amygdala—choose the response that is believed will
best ensure your survival. The survival responses include fight, when you attack or lash out at
the threat in an attempt to scare or even obliterate it; flight, when you run from the threat; and
freeze, when you stay very still to avoid attention by the threat. Which survival response the
brain chooses depends on several factors, including (but not limited to) training (such as mili-
tary training or personal defense classes), past traumatic experiences, social norms (which
might say that anger in women is not acceptable), or personality/temperament. Regardless of
which response your brain chooses during a stressful or traumatic event, it is sometimes helpful
to remember that your brain is trying to keep you safe with the response it has selected.

When a person repeatedly expresses unhealthy or disproportionate anger after experiencing
trauma, it is often conceptualized as the person entering the "fight" survival response, but in a
way that is unhelpful to the situation. While your brain's goal is to help keep you alive, it can
miscalculate threat after trauma, responding to nonthreatening situations or nonurgent stressors as though they are life-threatening. This can be problematic and damaging to your life and

relationships if you impulsively respond to loved ones, friends, or coworkers as though they are dangerous when they aren't.

Sometimes clients ask, "Why do trauma survivors go into 'fight' mode instead of 'flight' or 'freeze'?" The answer, first of all, is that not all survivors do. Instead of going into the "fight" survival mode, some survivors freeze or flee in response to stress after experiencing trauma.

People who tend to go into "fight" mode after experiencing trauma usually do so for one of two reasons: It was the response their brain selected at the time of the trauma or it is the response they *wish* they could have had at the time of the trauma. In the first scenario, you can think of the fight response as a sort of "brain habit" that forms after a traumatic event, where the brain learns that "fight" kept you alive. Because the brain remembers that the fight response kept you alive during the trauma, it reasons that it will keep you alive when you're faced with future stressors as well. The brain isn't wrong: people tend to steer clear of very angry people and avoid them. But when the goal is to enjoy life and connect with others, and when there is usually very little danger present in a person's environment, this knee-jerk survival reaction can be destructive.

You also may go into "fight" mode if you experienced a freeze response at the time of a trauma that left you feeling helpless. In John's case, he experienced helplessness as a child when his father terrorized the family. Sometimes you may remind yourself that you could not take action to stop the original trauma, and understandably, you never want to feel like that again. In order to feel powerful, you may lash out when faced with stressors. In a sense, it's an attempt to correct, or repair, the feelings of helplessness you experienced during the trauma. Ironically, however, while the goal is to become aggressive to gain a sense of power and control, dysregulated anger often produces the opposite experience, where you once again feel ashamed and out of control.

To better understand how dysregulated anger develops after trauma, here is a summary of what many people experience.

1. **You encounter a trauma reminder or perceive a threat/danger, and the amygdala activates.** This amygdala activation is your brain's way of warning you that a threat is present. Sometimes after trauma, the amygdala becomes not only hyperactive (overactive) but also hyper*reactive*, where it overreacts in different situations. While amygdala activation, or reactivity, can be experienced as fear, leading to a

"flight" survival response, it may also be experienced as a "fight" response. If you've been unable to keep yourself safe in the past, your brain might choose to interpret this amygdala activation as anger, as a way of empowering you to take action to protect yourself. This may especially be the case if past fear responses have led to additional harm being done to you. Thus, the amygdala may decide that the best way to keep you safe and alive is to lash out against the perceived threat, much like it did in John's case.

2. **Your "logical brain" goes offline.** When the amygdala activates, one of the consequences is that the rational, logical area of your brain, the prefrontal cortex, deactivates and turns off. This makes it very difficult for you to reason with yourself and "catch" your anger before you act out in a way you might later regret. With the survival brain (amygdala) in full force and your rational brain (prefrontal cortex) offline, you may feel as though you've "lost your mind," which is true, as you can't easily pause and analyze the situation to determine whether your reaction is disproportionate.

3. **Your behavior is impulsive and sometimes regrettable.** When you've "lost your mind," it can lead to behavior that you feel guilty or embarrassed about after the fact, which is how John felt every time he blew up at his family. When this happens repeatedly, some people feel as though they can't trust themselves or their reactions, which makes them want to isolate or avoid situations that might be "triggering" for them.

Use the following exercise to reflect on how you express anger. You can also visit http://www.newharbinger.com/48848 to download a copy of this exercise.

How You Express Anger

Over time, we tend to form habits about how we express our anger. Sometimes, when anger feels out of control, we develop behavior patterns, or habits, that don't serve us or those around us. Consider the following as you reflect on the ways in which you tend to express anger.

Do you tend to feel guilty after becoming angry? If so, is it the emotion itself you feel guilty about (meaning, do you feel guilty that you *felt angry*?) or is it the way you expressed it that makes you feel regret? Why?

_____

_____

_____

_____

When you're angry, do you often feel out of control? What's that like? Are there times you *are* in control while angry? When? How is that experience different?

_____

_____

_____

_____

What are your own triggers for "losing your mind"? Do you find yourself avoiding those triggers? Are there costs or consequences to doing that (such as missing out on fun events, time with family, etc.)?

_____

_____

_____

_____

If you find that you feel out of control while angry, experience frequent regret or guilt, and miss out on life as a result of dysregulated anger, this may be an area for you to focus on with a therapist.

Terms Your Therapist Might Use to Explain This Challenge

If your experiences are similar to John's, your therapist may use the following descriptors and terms in your discussions, or in your therapy notes:

- **Arousal and reactivity symptoms.** This is one of the symptom clusters of PTSD according to the *DSM-5* and is comprised of six symptoms, including irritability/aggression, destructive behavior, hypervigilance, sensitive startle response, difficulty concentrating, and difficulty sleeping. Although trauma survivors often do not experience all of these symptoms, it is common for individuals to report one or more of them.

- **Emotion dysregulation.** This does not refer to a specific disorder but is a general way of conveying that a person has difficulty managing overwhelming, presumably negative, emotions. Usually, anger is the main emotion that individuals have the most difficulty regulating, and this is seen across many categories of disorders, including trauma- and stressor-related disorders, anxiety disorders, and mood disorders.

What Your Brain Needs

The amygdala and prefrontal cortex are two areas of the brain that can be shifted in the direction of health. These changes can be elicited through psychotherapy, medication, relaxation practices, body-based exercises, and cognitive behavioral techniques. If you've noticed that since your trauma your experience of, and reaction to, anger has changed, take heart that your experience can change *again*, and your brain can learn how to be less reactive to triggers and stress. Here is what your brain needs for you to better regulate anger.

1. Bottom-up techniques that downregulate the "threat detection center" of the brain (amygdala). Dysregulated anger is often a result of misinterpreting situations as

being more threatening than they actually are, and this happens when the amygdala is hyperactive or hyperreactive. While some people may interpret this amygdala activation as fear (as described in other chapters), it's sometimes the case that amygdala activation is experienced as anger instead. Just as bottom-up techniques help reduce and manage fear through regulating the amygdala, they can do the same for anger as well.

2. Top-down techniques that reduce reactivity and activate the "thinking center" of the brain (prefrontal cortex). Because anger reactions can occur quickly, it can be helpful to be able to pause and think before reacting, as this reduces the chances of behaving in a regrettable way. Top-down techniques strengthen the brain region that allows you to do just this—pause and reflect—so that you can catch your anger, process it, and refrain from acting on it impulsively.

The next section steps you through five skills that can help you learn how to recognize and regulate anger.

Skills for Meeting the Challenge

This section focuses on anger management skills that can help you downregulate the amygdala (threat detection center) and strengthen the prefrontal cortex (thinking center). These skills include a progressive muscle relaxation exercise, a breathing and visualization technique, a strategy for checking in with yourself about your anger intensity, and two cognitive techniques for regulating anger.

Skill 1: Tension Release (Bottom-Up)

Muscle tension is commonly associated with the stress response in the body (HPA axis activation) and brain (amygdala activation). For some, the stress response is interpreted as fear, and there is a tendency to pull away or shut down. For others, the stress response is associated with anger, and an urge to take action and "fight" the threat the brain has detected. When this occurs, muscle tension is common, as muscle tension prepares you to fight (or flee). One way to manage the stress response, and angry reactions, is to relax your muscles, which sends the message to your brain to relax.

One method of reducing muscle tension is by practicing a bottom-up, self-awareness-building technique called tension release, a form of progressive muscle relaxation (PMR). In PMR, you intentionally tense your muscles, and then release that tension as much as possible. As you practice this technique, you'll learn how to detect subtle signs of stress and tension in the body and release muscle tension on command. Just as with other techniques, practice makes progress. With practice, you will gain an improved sense of self-awareness and will begin to notice early signs of anger by paying attention to what your muscles are telling you.

During tension release, you will squeeze and then relax several muscles that tend to tense when a person feels angry. To begin this practice, close your eyes as you assume a comfortable position, either lying down or sitting up. Gently shift your focus to the breath, checking in with the breath and noticing whether your breathing is fast or slow, deep or shallow. Next, follow these instructions:

1. **Legs:** Begin to feel into your legs, noticing any sensations, distress, or tension in this region. Now, gently tense your leg muscles at approximately 50 percent intensity, and hold this tension for a few seconds. You may do this by extending your legs outward if you are sitting in a chair. Finally, relax your legs completely, noticing what it feels like to relax the legs after being tense. What does this relaxation feel like? How does it feel different from tension? Feel into this area for a few more seconds before moving on to the next area.

2. **Buttocks:** Begin to notice your buttocks, becoming aware of any tension in this region. Now, gently tense these muscles at approximately 50 percent intensity, and hold this tension for a few seconds. Notice what it feels like for this region to be tense. Finally, relax your buttocks completely, noticing what it feels like to be relaxed after having been tense. What does this relaxation feel like? How does it feel different from tension? Feel into this area for a few more seconds before moving on to the next area.

3. **Abdomen:** Begin to notice your abdominal muscles, becoming aware of any tension in this region. Now, gently tense your stomach muscles at approximately 50 percent intensity, and hold this tension for a few seconds. Notice what it feels like for this region to be tense, and how it might restrict your breathing. Finally, relax your abdominal muscles completely, noticing what it feels like to be relaxed after having been tense. What does this relaxation feel like? How does your breathing change?

How does it feel different from tension? Feel into this area for a few more seconds before moving on to the next area.

4. **Hands:** Begin to notice your hands and the muscles in your hands, becoming aware of any tension in this region. Now, gently tense your hands at approximately 50 percent intensity, and hold this tension for a few seconds. You may do this by making fists with your hands. Notice what it feels like for your hands to be tense. Finally, relax each hand completely, noticing what it feels like to be relaxed after having been tense. What does this relaxation feel like? How does it feel different from tension? Feel into this area for a few more seconds before moving on to the next area.

5. **Arms:** Begin to notice the muscles in your arms, becoming aware of any tension in this region. Now, gently tense your arm muscles at approximately 50 percent intensity, and hold this tension for a few seconds. You may do this by extending your arms outward and squeezing them. Notice what it feels like for your arms to be tense. Finally, relax your arms completely, noticing what it feels like to be relaxed after having been tense. What does this relaxation feel like? How does it feel different from tension? Feel into this area for a few more seconds before moving on to the next area.

6. **Upper back:** Begin to notice the muscles in your upper back, becoming aware of any tension in this region. Now, gently tense your upper back at approximately 50 percent intensity, and hold this tension for a few seconds. You may do this by pulling your shoulders back and imagining that you are trying to make your shoulder blades touch. Holding this tension, notice what it feels like for your back to be tense. Finally, relax your back completely, no longer squeezing your shoulder blades together, and notice what it feels like to be relaxed after having been tense. What does this relaxation feel like? How does it feel different from tension? Feel into this area for a few more seconds before moving on to the next area.

7. **Shoulders:** Begin to notice the tops of your shoulders and the muscles in this area, becoming aware of any tension in this region. Now, gently tense your shoulders at approximately 50 percent intensity, and hold this tension for a few seconds. You may do this by gently lifting your shoulders toward your ears. Notice what it feels like for

your shoulders to be tense. Finally, relax your shoulders completely, allowing each to lower, noticing what it feels like to be relaxed after having been tense. What does this relaxation feel like? How does it feel different from tension? Feel into this area for a few more seconds before moving on to the next area.

8. **Neck:** Begin to notice the muscles in your neck, becoming aware of any tension in this region. Now, gently tense your neck at approximately 50 percent intensity, and hold this tension for a few seconds. You may do this by gently lifting your chin. Notice what it feels like for your neck to be tense. Finally, relax your neck completely, noticing what it feels like to be relaxed after having been tense. What does this relaxation feel like? How does it feel different from tension? Feel into this area for a few more seconds before moving on to the next area.

9. **Jaw:** Begin to notice your jaw muscles, becoming aware of any tension in this region. Now, gently tense your jaw at approximately 50 percent intensity, and hold this tension for a few seconds. Notice what it feels like for your jaw to be tense. Finally, relax your jaw completely, noticing what it feels like to be relaxed after having been tense. What does this relaxation feel like? How does it feel different from tension? Feel into this area for a few more seconds before moving on to the next area.

10. **Face:** Begin to notice the muscles in your face, becoming aware of any tension in this region. Now, gently tense your facial muscles at approximately 50 percent intensity, and hold this tension for a few seconds. You may do this by imagining that you are scrunching your face like a crumpled up piece of paper. Notice what it feels like for your facial muscles to be tense. Finally, relax your face completely, noticing what it feels like to be relaxed after having been tense. What does this relaxation feel like? How does it feel different from tension? Feel into this area for a few more seconds.

Now, with your whole body relaxed, take a moment to just notice what your body feels like when tension is not present. You may begin to open your eyes as you end this practice.

TIPS FOR TENSION RELEASE

As you practice this exercise more over time, you'll begin to notice subtle indications of tension in the body faster, which will alert you to early signs of stress and anger. When you can notice these early signs of tension, you may intervene with tension release or other exercises that relieve the tension in your body associated with anger. Thus, it is best to practice this exercise multiple times per week, for a few minutes per day, if your goal is to become more aware of early signs of anger in the body. If there is an area of the body that experiences pain, you may tense that area less, at about 10 percent, or not at all, and instead just feel into each area.

Skill 2: Exhaling Anger (Bottom-Up)

Deep, slow breathing is a great, fast way to reduce activation of the "threat detection center" of the brain (amygdala). In this exercise, deep breathing is combined with a body scan and guided imagery to help you feel calmer and more grounded during moments of frustration and upheaval. Follow these instructions:

1. Begin by gently closing your eyes, if you'd like to, and bringing your attention inward to the breath. Just notice the quality of the breath, and what it feels like to inhale and exhale.

2. Now, begin to extend and elongate the breath, breathing fully in as you inhale, and breathing fully out as you slowly exhale. Set your intention to breathe in this manner.

3. As you continue to breathe deeply, bring your attention to a question: "Where is the anger in my body?" Sit for a moment with this question and see whether you notice anywhere in the body where the anger may be present.

4. Next, maintaining focus on this question, start at the bottom of your body, at your feet, and slowly move your awareness upward, through your legs, midsection, arms, stomach, chest, back, shoulders, neck, and head, noticing whether your anger is present in any of these regions. Make a mental note of where you experience anger in your body.

5. Continuing to breathe deeply, imagine in your mind's eye an image of your body, with the regions holding anger appearing in one color. The anger may be any color you choose. Sit for a moment, visualizing your body and the "anger color" being present in the regions you've associated with your anger.

6. Now, shift your attention to the areas of your body that do *not* contain anger, and begin to imagine a color that symbolizes these peaceful areas. This may be called a "peaceful color." Identify your peaceful color.

7. As you continue to breathe deeply, now imagine in your mind's eye that as you inhale, you breathe in your peaceful color, watching it spread throughout your whole body. As you exhale, imagine that you are breathing out your anger color, watching it leave your body. Stay here for a few moments, inhaling peace and exhaling anger.

TIPS FOR EXHALING ANGER

Breathwork, broadly, can be very helpful for regulating anger. This technique in particular can be useful when you experience anger that you are not able to take action on, or anger that would be unwise to act on. When you can't take action externally, it can benefit you to regulate yourself internally, to reduce stress and feelings of overwhelm. Do this technique as a regular practice, for five to ten minutes, to calm anger and stress.

Skill 3: Monitoring Anger (Top-Down and Bottom-Up)

One reason that anger can become difficult to manage is that it can sneak up on you and seem to overcome you with little warning. However, if you can better monitor your anger, you can better manage it. One way to do this is to frequently check in with yourself about your anger level using what therapists call the "subjective units of distress scale" (SUDS). Taking your own "SUDS" is a fast way to keep track of your emotions (including, but not limited to, anger) so that you can detect anger and address it before it becomes too intense and overwhelming.

However, oftentimes anger sneaks up on us, blindsiding us. Not knowing when anger might appear can make it difficult to monitor and manage, but for most of us, there are trends we can pay attention to and triggers that may be predictable. Complete the following worksheet to gain an understanding of when and why your own anger may arise, and you'll be able to better anticipate your emotional reactions. Then, once you can predict and detect the presence of anger, you can take your SUDS. You can also visit http://www.newharbinger.com/48848 to download a copy of this exercise.

Predicting Anger Worksheet

Answer the following questions to become more aware of situations and times when your anger may flare.

Situations when I tend to become irritated, frustrated, or angry:

_____

_____

Times of day when I tend to become irritated, frustrated, or angry:

_____

_____

People who tend to make me feel irritated, frustrated, or angry:

_____

_____

The last five times I can remember being angry (the situations, what was happening, and who was there):

_____

_____

Some patterns I can identify from the situations above (What do the situations have in common? Do they all involve work, or a particular person? This may help you notice trends, or triggers, with regard to your anger):

_____

_____

When you are able to anticipate your anger before it develops, you can take faster, more effective action in monitoring and managing it. Next, you take your SUDS, as described in the next section, whenever you find yourself in, or approaching, a situation (or person, etc.) that tends to elicit frustration or anger. If you find that you're frequently angry and are not sure what might be triggering it, consider setting an alarm on your phone to check in with your anger multiple times per day. Over time, you'll begin to better anticipate when and why you might become angry.

HOW TO MONITOR YOUR ANGER WITH SUDS

Taking your SUDS (subjective units of distress scale) is a great way to monitor your anger. Here's how:

1. **Check in with yourself about physical sensations occurring.** Do you notice anything happening in your body right now? If so, what? Are the sensations you notice ones that you associate with emotion? Do you detect anger in your body right now? You may specifically feel into your body to notice any muscle tension (including jaw tension), shallow breathing, fast heart rate, or feelings of being warm or hot, as these are a few indicators of anger for some people.

2. **Check in with yourself about thoughts.** What have you been thinking about over the past several minutes, or even couple of hours? Has your mind become stuck on one topic? Are you replaying something from the past, or rehearsing something future-based? Is it distressing or angering to you?

3. **Check in with yourself about your behaviors.** Over the past several minutes, or even couple of hours, have you been *doing* anything that suggests you may be experiencing anger? For instance, some people will fidget, pace, clean, organize, eat, or engage in other behaviors when they feel angry or upset. While these behaviors do not necessarily imply anger, over-engaging in cleaning, for example, may suggest you are attempting to regulate a distressing emotion.

4. **Take your SUDS.** Now, taking into consideration what you've noticed about your physical sensations, thoughts, and behaviors, how angry do you feel, on a scale of 1 to 100, where 1 is no anger at all, and 100 is the most anger you can ever imagine experiencing? This is your current SUDS.

5. **Reflect on what's happening.** Is it clear why your anger might be present, if you're noticing it? Has something happened today, or recently, that has led to an increase? Sometimes anger increases are linked to things that are happening around us, but sometimes they're a result of our own thoughts (about past events, etc.). Check in with yourself, reflecting on what's been happening in your life. Can you identify a source of your current anger?

TIPS FOR MONITORING ANGER

If you experience difficulties with anger on a regular basis, you may consider checking your SUDS multiple times each day, perhaps even setting a phone alarm to remind you to check in as described in the last section. Additionally, if you know what your anger triggers tend to be, be sure to monitor your anger and take your SUDS when you encounter those triggers (or anticipate that you might be faced with them). When you can catch ruminative thoughts and physical sensations of stress that might be elevating your anger, it gives you the opportunity to intervene and practice skills (such as relaxation techniques) that can help you downregulate and better manage your anger.

Skill 4: Rethinking Anger (Top-Down)

Part of what can make anger (or other emotions) so distressing is our interpretation, or appraisal, of it. For instance, when you think of your anger as dangerous, bad, or prohibited, it can make you feel worse, and may actually intensify the anger instead of reducing it. Additionally, our interpretations about situations strongly shape our emotional reactions to them and whether we feel anger. While we often think that specific events or situations (or people!) directly cause our emotional reactions, this is not the case; situations themselves do not "make" us feel bad or good. Rather, it is how we think about them that creates suffering. Rethinking anger is a mental exercise that helps you evaluate and think differently about the events and situations that have brought up anger in you, so that you may be able to better regulate this emotion. To practice rethinking anger, complete the following worksheet. You can also visit http://www.newharbinger.com/48848 to download a copy of this exercise.

Rethinking Anger Worksheet

When you can think differently about a situation, you may feel differently about it as well. Complete this worksheet to see if there are additional interpretations of events that have sparked your anger.

Bring to mind, and name, a situation or an event that made you feel angry. Ideally, this is something that is still unresolved for you or that you still feel angry about when you think about it. (This should not be a traumatic event, however.)

Name of the situation: _____

Next, describe the event or situation in a few sentences. Note what happened in the event, who was involved, where it was, and what happened. You can also include this information in the chart at the end of this exercise.

_____

_____

_____

Now, thinking about the event, ask yourself, "What made me angry about this? What piece of it, or what was it that made me angry?" Reflect for a moment on what, specifically, about the situation is angering to you. If you had to put it in a sentence, what would it be? Take a moment to write it down.

What I am angry about: *Example:* My friend walked by me in the hall and didn't acknowledge me because she's selfish and doesn't care about anyone but herself.

What I am angry about: _____

_____

Rate your SUDS score when you think about this interpretation.

My SUDS score: _____

Looking at your sentence, it likely reflects a particular interpretation about the situation that occurred. While it may not be incorrect, or inaccurate, there may be other ways of viewing the situation that are *also* accurate. Take a moment to jot down other interpretations of the situation that may contain some amount of truth. It's okay if the alternative interpretations don't feel totally true; just think about other ways of looking at the situation that may *contain some truth*. As you do this, the goal is to identify other interpretations of the situation that elicit *less anger* than the one you initially identified.

Other interpretations: *Example:* My friend didn't smile at me because she was lost in thought, not because she was trying to ignore me.

Other interpretations: _____

_____

Notice how you feel as you read your alternative explanations. Even if the interpretations are not completely true, you may notice that you experience less anger when you consider them, even if that just means adding them to your original interpretation (instead of completely replacing your original thought with these new ones). Rate your SUDS score when you think about this alternative interpretation.

My SUDS score: _____

You can use the following chart to consider and rate multiple situations.

Event/Situation	Interpretation	How Angry I Feel When I Consider this Interpretation (0-100 SUDS)	Alternative Interpretation	How True This Interpretation Likely is (0-100 Percent True)	How Angry I Feel When I Consider This Interpretation (0-100 SUDS)
Example: My friend passed me and didn't smile or say hello.	She doesn't care about me or is mad at me.	50	She didn't smile at me because she was lost in thought, not because she was trying to ignore me.	25 percent	10

TIPS FOR RETHINKING ANGER

This can be a useful exercise whenever you find yourself thinking about something angering, and there may be some flexibility in your interpretation of it. This may not apply to all situations, but in many contexts, the truth is more complex than we may initially consider. Learning how to reframe, and reinterpret, events can help us better manage and downregulate anger. Complete this exercise whenever you notice that you're feeling angry about something that has happened. You may also complete it for past events that you remember feeling angry about, especially if you can still access that anger now. While anger can be useful and can help us right the wrongs in the world, sometimes it's less helpful than we'd like it to be.

Skill 5: Anger Management Training (Top-Down and Bottom-Up)

One way to learn how to better manage anger is to practice getting angry! While this may seem counterintuitive (most people want to feel less anger, not more), it can actually be helpful to learn how to experience, and then reduce, low-level anger. When you learn how to downregulate mild anger, that skill can help you manage more intense anger as well. To practice anger management training, follow these steps:

1. **Remember something that made you feel angry.** Begin by bringing to mind a memory where you experienced anger. It is okay if this issue has been resolved. The goal here is to recall a past event that made you angry, and in doing so, experience a bit of that anger now, in real time. If possible, focus on a memory that can produce angry feelings of an intensity level of around 30 (or higher) on a scale of 1 to 100, where 1 is no anger and 100 is the most anger you can imagine experiencing (see skill 3, Monitoring Anger, for more information on rating and monitoring anger).

2. **Connect with the memory.** Next, try to fully connect with this memory. You may close your eyes, if that is helpful. As much as possible, try to recall what happened, including who was involved, where you were, and any sensory experiences you might have had at that time. Recall the memory in as much detail as you can.

3. **Notice how you feel now.** Notice how you feel about the memory now, as you recall it. Notice what the anger feels like in your body and any associated thoughts. If possible, fully connect with this anger, and once more rate it on a scale of 1 to 100.

4. **Focus on the breath.** Now, begin to disengage from the memory, no longer focusing on it. Instead, shift your focus to the breath, beginning to breathe in and out, slowly and deeply. You may also practice skill 2, Exhaling Anger, as a way to relax and reduce feelings of anger in this exercise. Continue to focus just on your breath, flowing and extending your breathing, for five to ten minutes.

5. **Notice your anger.** Finally, focus once more on your anger, briefly checking in with yourself about your anger. Now that you've been breathing, how intense is your anger, on that scale from 1 to 100? You will likely notice that your anger score has dropped as you've calmed your body and brain.

TIPS FOR ANGER MANAGEMENT TRAINING

Practice this exercise when you're feeling relatively calm, not in the throes of anger, as it is meant to briefly increase anger. Practicing how to be angry, and how to regulate it, can be a helpful skill to build and can prepare you to better manage more intense anger as it arises.

A Pause to Reflect

Now that you've come to the end of this chapter, take a moment to pause and reflect on the anger management skills and exercises that have been most helpful to you. This list, which you can fill in using the following chart, can help you remember what to focus on, and what to practice, when you're feeling angry and want to better manage that anger. You can also visit http://www.newharbinger.com/48848 to download a copy of this exercise.

Skills for Managing Anger

The following chart is a reminder of the skills in this chapter and a place to record how each skill works for you. When you notice anger is negatively impacting your life, choose a skill that is appropriate for the situation. The second column tells you when the best time to practice is. In the third column, make a note of when and how you try the skill over the coming week. Then, take a moment to reflect on your experience. If you have learned other useful skills or techniques that help, add them to the chart in the blank spaces. If you need more space, you can print more pages or keep a journal to reflect on your experiences.

Skill/Technique	Best Time to Practice	When I Practiced the Skill (Day and Time as Well as Description of the Situation)	Reflect on the Experience. Was the Skill Useful?
Tension Release	As a regular practice or when I feel angry.		
Exhaling Anger	When I feel angry.		
Monitoring Anger	When I feel angry.		
Rethinking Anger	When I feel angry.		
Anger Management Training	As a regular practice or when I feel angry.		

CHAPTER 4

Strong, Distressing Physical Reactions

Peter, a forty-five-year-old man, has been going to therapy for more than six months, but doesn't feel like he's making progress. He began therapy after his wife expressed concern that he "didn't seem like [himself]" after surviving a terrible car accident. Peter says that when his therapist asks him about distressing topics, including the accident, his body reacts so strongly that he finds it difficult to focus on answering her questions. As these sensations intensify he feels increasingly out of control, and worries that if he opens "Pandora's box" he will feel even worse. Peter realizes that he needs to process what happened to him, but he feels sick and claustrophobic every time he tries to talk about it.

Check-In: Your Experiences

After reading about Peter, take a moment to think about your experiences. You may choose to reflect and journal on the following questions:

- Do you notice anything upsetting or distressing that happens in your body when you think about, or are reminded of, your trauma(s)? If so, what do you notice?

- How do these sensations impact your life? For some people, the sensations lead them to avoid people, places, or situations. Do your distressing physical reactions influence or alter your life in any way?

How Distressing Physical Reactions May Impact Your Life

Distressing physical sensations are very common after trauma and are one of the most pressing complaints due to their intensity, frequency, and duration. These sensations can be extremely aversive, which motivates you to try almost anything to get them to stop! Often, when you feel these sensations you have difficulty thinking clearly or focusing, as trauma reactions in the body can become all-encompassing and overwhelming. You may also report feeling panicked, enraged, or disconnected from yourself or your environment when these sensations become strong. This can lead you to avoid people, places, or situations that elicit those sensations, which can negatively impact your relationships, work or school life, and overall well-being.

The following brief self-assessment can help you determine whether strong, distressing physical reactions are causing you problems in your life. You can also visit http://www.newharbinger .com/48848 to download a copy of this exercise.

Distressing Physical Reactions Self-Assessment

Read the following and circle the best response from 0 to 3, where 0 = None/not at all, 1 = A little bit/sometimes, 2 = A moderate amount/often, and 3 = A lot/most of the time.

When something reminds me of the trauma or when I think or talk about it, I experience…	0	1	2	3
Chest tightening	0	1	2	3
Throat constriction	0	1	2	3
Pressure in the head or headache	0	1	2	3
Feeling hot	0	1	2	3
Dizziness/light-headedness	0	1	2	3
Shallow breathing	0	1	2	3
Rapid heart rate	0	1	2	3
Clammy palms	0	1	2	3
Nausea	0	1	2	3
Muscle tension, especially in the neck, shoulders, back, jaw, and hands	0	1	2	3
Tingling	0	1	2	3
"Waves" of nausea, tingling, temperature changes, or energy that moves upward or downward, especially in the stomach or back	0	1	2	3
That I avoid some situations or places because they might elicit distressing sensations.	0	1	2	3
That I avoid some people because they might elicit distressing sensations.	0	1	2	3
That distressing sensations tend to make it difficult to sleep.	0	1	2	3
That I lose my appetite because of distressing sensations.	0	1	2	3
That I feel overly aware of my distressing sensations.	0	1	2	3
That sometimes I drink or use drugs to calm my distressing sensations.	0	1	2	3
Total				

Add up the total score from the above items, and record it next to "Total." A score of 16 or higher indicates that trauma-related physical reactions may be causing you difficulties in your life.

Use the following exercise to reflect on your self-assessment score. You can also visit http://www.newharbinger.com/48848 to download a copy of this exercise.

Reflect on Your Self-Assessment Score

What did you score on the self-assessment? Did you learn anything new? Take a moment to consider the following questions.

Was your score higher than you thought it would be? Lower?

_____

_____

Were there any sensations you didn't know you were experiencing so strongly?

_____

_____

_____

Were there items on the assessment you hadn't considered before?

_____

_____

_____

Did you learn anything about how your distressing reactions might be impacting your life?

_____

_____

_____

Why Distressing Physical Reactions After Trauma Are So Common

The physical reactions listed in the self-assessment are some of the most common ones that trauma survivors report. What you may not know is that all of the physical reactions listed here are indicative of a strong stress response in the body and brain. While we all experience physical stress reactions sometimes, after trauma these sensations can intensify and feel like an amped-up, magnified version of "normal" stress that is difficult to tolerate. Ultimately, the purpose of such strong sensations is to keep you safe, but the way the body and brain do this is to actually first convince you that you *aren't safe*, which motivates you to take some sort of action to remove yourself from danger. Here's a rundown on what's happening when these sensations arise.

1. **You get "triggered."** First, there is a "trigger" of some sort. This might be a thought about a traumatic event that suddenly pops into your mind, a trauma reminder you see on television or somewhere else, or a comment someone makes that activates a trauma memory. Triggers can also be subtle. For instance, a change in the weather may be a trigger, or the introduction of holiday decorations in stores, or a certain smell that you don't consciously associate with anything negative. Some triggers seem to be linked to particular times of year or recur predictably every few weeks or months. They can be internal (arising from your own thoughts, for example) or external (arising from the environment or other people) and may be obvious or elusive.

2. **Your brain's "smoke alarm" goes off.** When confronted with a trauma reminder, whether subtle or strong, your brain immediately identifies it as a potential danger. Why? Because in the past when the trauma occurred, that reminder, which is now a trigger, was present. Even if the trigger had nothing to do with the event—such as the case with hot weather that you'd think to be irrelevant to a mugging—it may still set off the brain's alarm bells if it is associated with the trauma in any way. Someone who was mugged on a hot summer night may later feel triggered by hot or humid weather, in addition to several other environmental cues such as nighttime, parking lots, people who look like the mugger, and the sound of footsteps. When this happens, the amygdala, the area of the brain responsible for detecting threat

and danger, activates. This in turn sets into motion a series of reactions that produce distressing sensations in the body.

3. **Your stress pathway activates.** After the smoke alarm (amygdala) activates, it sends threat signals to other areas of the brain that help you prepare to deal with the perceived (but possibly not real!) danger at hand. One area of the brain that receives the amygdala's danger signals is the hypothalamus. When this area of the brain gets word that you are in danger, it activates and spreads the threat signal to the pituitary, also in your brain. Together, the hypothalamus and pituitary, which are connected to one another, comprise the stress pathway in your brain. But interestingly, your stress pathway isn't just in your brain; it's in your body as well. Once the hypothalamus and pituitary are activated, thanks to the smoke alarm of the brain (amygdala), the threat signal is sent down to the adrenal glands, which are in your *body*. Together, the connection between the hypothalamus, pituitary, and adrenals—the HPA axis—is referred to as the "stress pathway," and it's physically located in *both* your brain and your body. This stress pathway is how the brain communicates danger and threat to the body.

4. **Your body goes into the stress response.** When the HPA axis activates, the danger signals move from your brain to your body. Once the signals reach your body, the sympathetic nervous system activates the stress response. The purpose of sympathetic arousal is to put your body into a state where it becomes faster and stronger. The logic is that if you are fast and strong, you are more likely to survive whatever danger is threatening you. In other words, you become stressed in order to stay alive!

5. **You experience distressing sensations.** When the body goes into the stress response, it doesn't usually feel good. You may begin to feel some of the sensations described earlier in this chapter. The sensations, being aversive, motivate you to take action to get out of the dangerous situation.

Use the following exercise to reflect on your distressing physical reactions. You can also visit http://www.newharbinger.com/48848 to download a copy of this exercise.

Your Distressing Physical Reactions

Now that you know more about how you experience distressing physical reactions, it can be helpful to think about when these tend to arise and which ones you might want to work on.

How often do you experience your most distressing physical reactions? Is it daily? Weekly? For how long, when they occur?

_____

_____

_____

Are the sensations predictable, or do they also sometimes catch you off guard? When do they tend to occur?

_____

_____

_____

_____

What do you do when you experience the sensations? Has anything ever helped ease them?

_____

_____

_____

_____

Terms Your Therapist Might Use to Explain this Challenge

If your experiences are similar to Peter's, your therapist may use the following descriptors and terms in your discussions, or in your therapy notes:

- **Physiological reactions to trauma reminders.** This usually just means that your body goes into a stress response when the brain detects trauma triggers in your environment (or your mind). The phrase "physiological reactions to trauma reminders" refers to a specific symptom that appears in the *DSM-5* as reexperiencing symptoms of PTSD.

- **Psychological distress in reaction to trauma reminders.** This usually refers to a general feeling of anxiety, sadness, or upset you might feel when faced with trauma triggers. While it isn't referring to physical sensations per se, physical sensations may contribute to these feelings. This, too, is a reexperiencing symptom of PTSD that appears in the *DSM-5*.

- **Arousal and reactivity symptoms.** This is one of the symptom clusters of PTSD, and is comprised of six symptoms: irritability/aggression, destructive behavior, hypervigilance, sensitive startle response, difficulty concentrating, and difficulty sleeping. This phrase does not directly refer to physical reactions, but those who experience strong, aversive, trauma-related sensations tend to report experiencing some of the symptoms listed here. For example, if you feel tense and experience rapid heart rate, you may naturally startle more easily than others and may feel on guard (hypervigilant) due to your brain and body telling you that you are in danger.

What Your Brain Needs

During times of intense physical triggering, two important things happen in the brain. First, as discussed in this chapter, the smoke alarm of the brain (amygdala) activates, setting off a chain of reactions that lead to the aversive sensations you feel. Second, the area of the brain that helps you feel into internal experiences, the "body sensor" (insula), becomes hyperreactive. When this happens, the physical sensations you feel become more intense, and you may have a difficult time focusing away from them, attending to other things, or managing them. So here is what your brain needs if you're experiencing aversive, trauma-related sensations.

- Bottom-up techniques that quiet the smoke detector (amygdala). If it's true that you're not actually in danger, and that the amygdala is needlessly activating in response to something that reminds it of a traumatic event, then a main goal is to practice techniques that train the amygdala to calm down and deactivate.

- Bottom-up techniques that help regulate the body sensor (insula). We want the insula to be online and activated, but not *hyperactive* or *hyperreactive*. Feeling safe, present, and grounded in the body, while also being able to access other experiences (such as thoughts), is what we're aiming for.

- Top-down techniques that help strengthen the mind (prefrontal cortex) and down-regulate the smoke detector (amygdala). Another way to reduce amygdala activation can be through accessing soothing thoughts or positive memories by activating the prefrontal cortex. Also, accessing these adaptive thoughts is a great way to strengthen the prefrontal cortex.

The next section details four skills that teach you how to regulate your brain's body sensor and calm the brain's smoke alarm, allowing you to better become aware of, tolerate, and manage sensations.

Skills for Meeting the Challenge

This section focuses on four tools and techniques to help you learn about and manage strong, distressing, post-trauma reactions in the body.

Skill 1: Distress in the Body (Bottom-Up)

It can be challenging to tolerate the physical sensations of trauma, as they are often experienced as painful and distressing. Trauma can change your relationship with your body, and this is especially true if trauma occurred to your body. In order to heal from trauma, it is important to be able to connect with physical sensations, even the distressing ones, in order to shift your relationship to them and manage them better. This exercise is a bottom-up, somatic technique where you learn how to notice internal experiences.

To notice and tolerate distress in your body, you first need to connect with it. At a time when you are not triggered, and your physical sensations are at "baseline," ask yourself the following questions, which can help you connect with the surface and internal sensations of your body.

SURFACE SENSATIONS

- **Airflow:** "Can I feel air on the surface of my body?" This might be wind while standing outside or where the air meets the skin while sitting still. For a moment, be still and notice any air current on the surface of your body.

- **Pressure:** "Can I feel any pressure on the surface of my body?" Starting at your feet, scan upward, noticing any pressure that may be present. It may be noticed, for instance, that there is pressure on your legs and backside if you are sitting down.

- **Sensations:** "Can I feel any sensations on the surface of my body?" External sensations may include tingling, itching, or other experiences.

- **Temperature:** "Can I feel the temperature of my body?" Starting at your feet, scan upward, noticing the temperature of various regions of your body and noting how they differ (for instance, your feet may be cooler than your stomach).

INTERNAL SENSATIONS

- **Airflow:** "Can I feel the air current flow into my body through my nose or mouth?" Instead of feeling the air on the surface of your skin, notice how the air feels as it enters your body through your mouth or nose. Notice where, inside your mouth or throat, the sensation of the air disappears and it flows down to your lungs.

- **Sensations:** "Can I feel any sensations on the inside of my body?" Internal sensations may include stomach rumbling, head pulsating, "butterflies," a buzzing feeling (which may indicate anxiety), or a pounding heart.

- **Temperature:** "Can I feel the temperature inside my body?" To notice internal temperature, consider focusing on your abdomen and inside your mouth, feeling into these areas and noticing sensations of warmth that might be present.

- **Tension:** "Can I detect any tension in my muscles?" Some areas of the body tend to hold tension, such as the neck, back, and jaw.

Next, "check in" with your body for the following sensations in order to detect physical indications of distress. If you answer "yes" to some of the following questions, it is likely that you are experiencing distress or anxiety, as all of the following signal stress in the body. If none of these sensations are present, you may also want to repeat this exercise when you've had a hard day or know you might be experiencing some stress. You can also visit http://www.newharbinger.com/48848 to download a copy of this exercise.

Physical Distress Questionnaire

1. Do I feel pressure in my head?

2. Do I feel pounding in my head?

3. Do I feel like there is a tight band around my head?

4. Does my head feel hot?

5. Does my face feel hot?

6. Do the muscles around my eyes feel tense?

7. Does my jaw feel tense or tight, like I am clenching my teeth?

8. Is there a ringing in my ears?

9. Does it feel difficult to swallow?

10. Does my neck feel stiff or sore?

11. Do my shoulders feel stiff or sore?

12. Does my back feel stiff or sore?

13. Does my chest feel tight?

14. Do I feel pressure in my chest?

15. Is my heart pounding or beating fast?

16. Does it feel like it is difficult to take deep breaths right now?

17. Is my breathing fast and shallow?

18. Does my stomach feel like it is "in knots"?

19. Does my stomach hurt?

20. Does my stomach feel like it is dropping?

21. Do my hands or feet feel cold?

22. Do my hands or feet feel shaky?

23. Does any part of my body feel shaky?

TIPS FOR DISTRESS IN THE BODY

When you experience some of the sensations that indicate physical distress, it may mean that your body has entered the stress response. The more you experience these sensations, the higher your distress may become. Try to practice this skill whenever you feel upset and distressed, to feel (in real time!) where the distress is located. Repeat this practice every two or three months as a sort of check-in; you may find that as you work on these symptoms, they change over time. Keeping a record of how they are shifting over time can be one way of gauging progress.

Skill 2: Creating a Physical Trigger Profile (Bottom-Up)

When you are under stress or feeling triggered, it can be helpful to refer back to the list of sensations outlined in skill 1, Distress in the Body, and "check in" with yourself, noticing and documenting which of the sensations you experience at those times. Once you identify sensation patterns, which are the sensations you consistently tend to experience when stressed or triggered, you can create a "trigger profile," which acts as a summary of your most commonly felt distressing sensations. Use the worksheet following the directions to map your own physical profile of emotions. You can also visit http://www.newharbinger.com/48848 to download a copy of this exercise.

1. Starting at the top of your body, identify and name the physical sensations you tend to experience when stressed or triggered. For example, you might feel a racing heart, shallow breathing, a knot in your stomach, and so on. Write the sensations on the lines to the right of the diagram.

2. Draw an arrow from the written sensation to the outline of the person, to indicate exactly where you experience the sensation. This will become your trigger profile, and it can help you anticipate how trauma reminders are likely to be represented in your body. This is the first step to learning how to better tolerate and manage distressing sensations.

Physical Trigger Profile

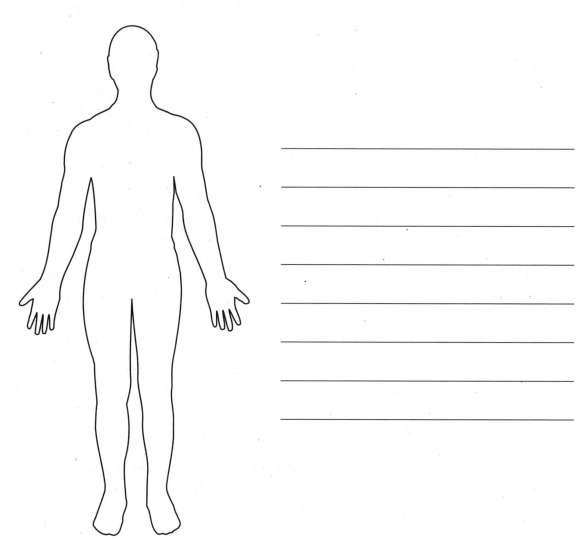

TIPS FOR CREATING A PHYSICAL TRIGGER PROFILE

This is not an exercise you'll complete often, because once it's complete, it will not likely change in the short term. However, it can be helpful to refer to this profile when you're feeling triggered, to verify the sensations and begin working directly with these areas of your body (refer to the next two techniques for ideas about how to manage distress in the body).

Skill 3: Identifying Safe Places Inside (Bottom-Up)

This technique helps you connect with your body and locate areas of calmness, comfort, or safety. When you can identify and connect with areas of your body in a positive way, they become "resourced," and you can connect with those areas during difficult times in order to feel a sense of strength or safety, and to reduce feelings of overwhelm.

1. Close your eyes, if you feel comfortable, or gently gaze downward at a spot on the floor. Focus your attention on your breath, breathing effortlessly, not trying to change your breath. Notice what it feels like as you breathe in and as you breathe out.

2. As you progress, you may notice various sensations. When this happens, simply experience and acknowledge the sensations, as well as any accompanying thoughts or emotions.

3. Shift your focus to the soles of your feet, just noticing, without judgment, any sensations that may arise as you attend to this area. Stay here for a moment and ask yourself, "Is there calmness, comfort, or safety in this area?" If the answer to your question is "no," begin to redirect your awareness to the next area of the body.

4. Gently shift your awareness to your toes and the top of your feet, noticing any sensations that are present, without judgment. Stay here for a moment and ask yourself, "Is there calmness, comfort, or safety in this area?" If the answer is "yes," remain focused here for a few moments, connecting more deeply with this area of your body and intentionally experiencing the strength this area is providing. Remember that you can connect with this resource whenever you need it.

5. Begin to redirect your awareness to the lower part of your legs, above the feet but below the knees. Notice any sensations in your calf muscles and in the front of your legs, without judgment. Stay here for a moment and ask yourself, "Is there calmness, comfort, or safety in this area?" If the answer is "no," begin to redirect your awareness to the next area of the body. If the answer is "yes," remain focused here for a few moments, connecting more deeply with this area of your body and intentionally experiencing the strength this area is providing.

6. Continue this exercise, moving your awareness through several major muscle groups and different regions of the body. Which particular areas you choose to focus on is up to you; the following areas can be scanned in this exercise:

   • Feet

   • Lower legs

   • Upper legs

   • Buttocks/hips/pelvic area (any or all of these areas)

   • Abdomen

   • Lower and/or upper back or entire back

   • Arms

   • Hands

   • Chest area (with a focus on the breath)

   • Shoulders

   • Neck

   • Head

   • Jaw

TIPS FOR IDENTIFYING SAFE PLACES INSIDE

This technique can become more helpful the more you practice it, similar to other mindfulness practices. Practice this technique on a regular basis as a way to calm, self-soothe, and resource.

Skill 4: Alternating Attention for Self-Soothing (Bottom-Up and Top-Down)

This technique trains you to do the following:

- Better tolerate and reduce distressing sensations

- Intensify positive sensations

- Direct and redirect your attention to different sensations

In skills 2 and 3 you learned how to pinpoint distressing and safe sensations in the body. This skill helps you learn how to work with *both* distressing and safe sensations and areas, desensitizing to distressing sensations and feeling more deeply into resourced areas. Desensitization is facilitated by tapping each shoulder, quickly, as you notice aversive sensations in the body; intensification of positive sensations and resourced areas is facilitated with slow, rhythmic tapping on your shoulders.

The reason why fast bilateral tapping (tapping each shoulder in sequence) facilitates desensitization of distressing material, including sensations, is that it subtly distracts you, pulling some of your attentional resources away from the distressing sensation(s) as you notice the shoulder tapping. Research has shown that fast bilateral stimulation, such as tapping, can dull upsetting and aversive experiences. However, very slow and rhythmic bilateral stimulation has been shown to intensify an experience, which is why it's best to *slowly* tap each shoulder when connecting with a resourced area of the body. The slow bilateral stimulation may make the resourced area of the body feel even stronger, calmer, or safer.

1. If you are currently feeling stressed or triggered, skip this step and move to the next one. If you are not currently experiencing any distressing sensations in the body, you might choose to elicit some of these sensations by remembering a time recently when you felt stressed, overwhelmed, or triggered. When you access this memory

you may notice that aversive sensations arise, which will allow you to fully practice this technique.

2. Close your eyes, if you are comfortable, or gently gaze downward at a spot on the floor. Begin to feel into an area of the body, or a sensation, that feels distressing to you (tight chest, racing heart, head pressure, etc.). This feeling may become strong when you feel triggered or overwhelmed or when you access an upsetting memory from the past. When you have connected with this distressing sensation, begin to gently but quickly tap each shoulder, in sequence. Continue focusing on this area as you quickly tap your shoulders for about twenty seconds.

3. After about twenty seconds of rapid tapping, stop tapping and begin to redirect your attention to an area of the body, or a sensation, that is free of any distress. This is a resourced area of the body that you may experience as strong, safe, neutral, or otherwise distress-free. While attending to this resourced area or sensation, tap your shoulders again, this time tapping each shoulder *very slowly* as you focus on the physical resource. Hold your attention in this area of your body for thirty to forty seconds.

4. After thirty to forty seconds of slow bilateral tapping, stop tapping and redirect your attention back to the distressing area or sensation. How does it feel now? Has anything changed about this sensation? Begin to rapidly tap your shoulders as you connect with the distressing sensation again. Hold your attention on this aversive sensation for another twenty seconds before stopping.

5. Continue to alternate your attention between the distressing sensation and the resourced area of the body as described earlier, incorporating fast tapping and slow tapping, respectively. To summarize, you will alternate between these two tasks:

    a. Focus on a distressing, stress-induced physical sensation (such as tight chest or nausea) while rapidly tapping each shoulder for about twenty seconds.

    b. Focus on a resourced area of the body that does *not* contain any distress (such as the feet, hands, or other safe area) while slowly tapping each shoulder for thirty to forty seconds.

6. Alternate between these two tasks for five to ten minutes, noticing as you alternate between the two how the distressing sensation might be shifting over time. Many people find that as they practice this skill their stress response lowers and the aversive sensation begins to subside a bit.

TIPS FOR ALTERNATING ATTENTION FOR SELF-SOOTHING

Try to complete this exercise with your eyes closed, if possible, to experience the technique more deeply. Also, adopt this as a regular practice so your body and brain can learn how to help you relax and desensitize to sensations. With a technique such as this one, practice makes progress; the more you practice, the better it tends to work.

A Pause to Reflect

Now that you've come to the end of this chapter, you may want to take a moment to pause and reflect on what you've discovered and the techniques that you've found helpful. These techniques might be ones described in this chapter or other chapters. The following worksheet can help you keep these techniques in mind for when you need them! You can also visit http://www.newharbinger.com/48848 to download a copy of this exercise.

Skills for Managing Distressing Physical Sensations

The following chart is a reminder of the skills in this chapter and a place to record how each skill works for you. When you notice distressing physical sensations are negatively impacting your life, choose a skill that is appropriate for the situation. The second column tells you when the best time to practice is. In the third column, make a note of when and how you try the skill over the coming week. Then, take a moment to reflect on your experience. If you have learned other useful skills or techniques that help, add them to the chart in the blank spaces. If you need more space, you can print more pages or keep a journal to reflect on your experiences.

Skill/Technique	Best Time to Practice	When I Practiced the Skill (Day and Time as Well as Description of the Situation)	Reflect on the Experience. Was the Skill Useful?
Distress in the Body	When feeling distressed, or when I can access thoughts about something that makes me feel distressed.		
Creating a Physical Trigger Profile	When feeling calm, to better understand triggers.		
Identifying Safe Places Inside	When feeling calm, as a regular practice, or when feeling upset.		
Alternating Attention for Self-Soothing	When feeling upset or triggered.		

CHAPTER 5

Emotional Numbing and Apathy

Claire, a thirty-eight-year-old woman, experienced ongoing childhood abuse and neglect while growing up. Although she feels proud of herself for how far she's come and the loving family life she has helped create, something seems to be missing. Claire says that she often feels as though she's going through the motions of life, unable to feel emotions, including love, excitement, and happiness, even toward her children. And even when she does experience emotion, she finds it difficult to describe and express those feelings. While she believes she "should" be happy, she has experienced difficulty connecting to her emotions for many years and describes her overall emotional state as numb and apathetic. She feels guilty about her lack of emotion and inability to laugh or cry, as it makes her feel like a failure as a mother and wife.

Check-In: Your Experiences

After reading about Claire, take a moment to think about your experiences. You may choose to reflect and journal on the following questions:

- Do you have difficulty accessing and feeling emotions, too, like Claire? For some people, emotional numbing is occasional, or fleeting, but for other people it can be more persistent and long term.

- Do you find it challenging to describe your feelings? Do you struggle to find ways to express your feelings?

- If you experience frequent numbing or apathy, what is that like for you? Do you find yourself wishing you could feel emotion, or does it feel safer to have emotions turned off?

How Feeling Apathetic May Impact Your Life

Many people don't realize that feeling numb or apathetic is very common after traumatic experiences. Additionally, many trauma survivors experience an inability to name, describe, and express emotion, and this is referred to as alexithymia. When you can't fully access love, joy, happiness, and contentment, as well as sadness, anger, and fear, life becomes dull and begins to lose meaning. Without emotion, it becomes difficult to connect with yourself and others, and this can adversely impact your relationships. Although some trauma survivors come to believe that feeling apathetic is "just the way they are now," this doesn't have to be the case. Oftentimes, after traumatic events your brain will shut off emotion as a way to self-protect. Trauma can produce overwhelmingly strong emotions that are very painful, and as a result, the brain may choose to disengage from emotion. This is especially the case if, after the trauma(s), no one was there to comfort you or meet your emotional needs.

The following brief self-assessment, which is adapted from the Toronto Alexithymia Scale and the Perth Alexithymia Questionnaire (Taylor et al. 1985; Preece et al. 2018), is a fast way to determine whether you might be experiencing apathy and alexithymia. You can also visit http://www.newharbinger.com/48848 to download a copy of this exercise.

Apathy and Alexithymia Self-Assessment

Read the following and circle the best response from 0 to 3, where 0 = None/never, 1 = A little bit/ sometimes, 2 = A moderate amount/often, and 3 = A lot/most of the time.

I have a hard time expressing my emotions.	0	1	2	3
I cannot put into words how I feel.	0	1	2	3
I feel confused by my feelings.	0	1	2	3
I find it difficult to connect with my emotions.	0	1	2	3
It is hard for me to feel close to other people emotionally.	0	1	2	3
People in my life complain that I don't show my feelings enough.	0	1	2	3
People in my life complain that they don't feel close to me.	0	1	2	3
Sensations in my body confuse me.	0	1	2	3
I don't know why I feel the way I do.	0	1	2	3
I avoid talking about my feelings.	0	1	2	3
I find it difficult to have conversations about other people's emotions or reactions.	0	1	2	3
I do better focusing on facts, rather than feelings.	0	1	2	3
I know when I'm feeling "bad," but cannot tell whether the emotion is fear, anger, sadness, or something else.	0	1	2	3
I know when I'm feeling "good," but cannot tell whether the emotion is happiness, joy, contentment, love, or something else.	0	1	2	3
I feel apathetic.	0	1	2	3
I avoid thinking about, naming, or describing my emotions.	0	1	2	3
I'm bothered by my lack of emotion.	0	1	2	3
I don't think feelings are as important as more tangible things, such as tasks.	0	1	2	3
It feels unnatural, forced, or awkward to think about my feelings.	0	1	2	3
I avoid noticing physical sensations that accompany some emotions (such as racing heart rate when fear is experienced).	0	1	2	3
I feel numb.	0	1	2	3
Total				

Add up the total score from the above items, and record them next to "Total." A score of 26 or higher indicates the presence of numbing, apathy, and alexithymia.

While most people experience some of these symptoms for short periods of time, when these symptoms persist, they can contribute to difficulties with self-awareness and relationships. Use the following exercise to reflect on your self-assessment score. You can also visit http://www.newharbinger.com/48848 to download a copy of this exercise.

Reflect on Your Self-Assessment Score

What did you score on the self-assessment? Was this surprising to you? Take a moment to reflect on, and process, what you just learned about yourself. You may consider the following questions as you reflect on your experience with this.

Was your score higher than you thought it would be? Lower?

_____

_____

Were there items on the assessment you hadn't considered before?

_____

_____

_____

How do you think these symptoms, or experiences, have been impacting your life, including your relationships?

_____

_____

_____

Have there been times in your life when you might have scored higher or lower on the questionnaire? How were those times different from your current life?

_____

_____

_____

Why Apathy and Alexithymia After Trauma Are So Common

Traumatic events produce extremely strong, distressing emotional reactions that often feel overwhelming to the survivor. These emotional reactions are not just cognitive—they aren't just happening in your head, with your thoughts. Rather, these reactions are most prominently experienced *in the body*, as physical sensations. In fact, what makes an emotion feel pleasant or distressing has almost nothing to do with thoughts; we *feel* emotions because we *feel* our bodies, and every emotion has a physical expression. While thoughts can lead to emotions, it's the physical sensations that produce the actual *experience* of the emotions.

When you experience a terrible, emotionally (and possibly physically) painful event, the body reacts strongly, usually catapulting you into a stress response (see chapter 4 for more information on this topic). After the event, it can happen that you reexperience those distressing sensations whenever you get stressed, reminded of the traumatic event, or for no apparent reason at all. This can lead to you feeling emotionally dysregulated and overwhelmed. However, there is another common reaction after trauma, which is the exact opposite of overwhelming emotion: apathy and numbing. Instead of reexperiencing the physical sensations of the trauma, you may instead notice that you feel *nothing* after experiencing traumatic events.

While this might sound like a welcome relief to survivors who are engulfed by their own emotions, feeling numb and apathetic can be distressing, as people often report that they feel ungrateful for the good things in their lives, disconnected from loved ones, and "cold." This can lead to self-blame, guilt, and shame about an emotional experience (apathy/numbing) that feels outside of your control. For example, some survivors say that they have a hard time accessing loving feelings toward their families, children, and friends. They may say, "I know I love my kids, but I just can't feel it, even when I'm with them." You can see how this can be very distressing, to know you love your family but not feel able to actually feel it!

It might also be the case that you can *feel* your emotions but cannot describe or express them. For instance, you may know that you feel "good" or "bad," but cannot name the specific emotion that is present or explain what the experience feels like. Thus, while some survivors experience overwhelming emotions after their trauma, others experience apathy, numbing, and alexithymia. Some people may experience both at different times, where they may feel rage in some moments and apathy in others.

Because apathy, numbing, and alexithymia can feel confusing and distressing, it can be helpful to understand what happens in the brain during these experiences and why they happen. One way to think about this is that after traumatic events, the brain may disconnect from emotions in order to protect you from further pain. This is primarily accomplished by turning off (reducing activation) of a key brain region involved in experiencing physical sensations and internal body experiences: the insula. Sometimes known as the "interoception center," where *interoception* refers to the ability to feel into the body and notice physical sensations, the insula (along with other brain areas) is tasked with helping you track, name, and describe what's happening in your body. But when you turn down the insula, dulling its activation, it becomes much harder to experience and understand your emotions, and you may feel numb, apathetic, and alexithymic.

To better understand how apathy and alexithymia develop after trauma, here is a summary of what many people experience.

1. **Trauma happens, and the amygdala activates, along with the insula.** When you experience trauma, your amygdala, which is the area of your brain responsible for detecting threat and danger, activates. This activation often occurs along with insula activation, which allows you to experience the stress response–based sensations in your body. When you feel these distressing sensations as a result of the trauma that's happening, it feels awful, and it motivates you to get out of the situation that is producing these terrible feelings. It's your brain's way of alerting you that something is terribly wrong and motivating you to get out of the situation and back to safety, whether that means fighting, fleeing, or freezing.

2. **After trauma, the insula can turn off.** In some people, the insula will deactivate, or turn off, after the traumatic event, as a result of it feeling so terrible. This is likely

to happen if, after the traumatic event, you did not receive support from caregivers, family, or friends. If the message was that you needed to "get over it" or "move on," then your brain may try to do this by cutting you off from your emotions, through cutting you off from your body.

3. **Low insula activation may produce apathy, numbing, and alexithymia.** When the insula is underactive, you may find it difficult to access, notice, or describe your feelings and emotions. This is in large part because your insula is the area of your brain that allows you to feel physical sensations, which is necessary for the experience of emotions (because, remember, emotions are largely experienced as physical sensations!). When the insula is not sufficiently activated, it becomes hard to experience, name, and understand emotion, because your ability to feel into your body and notice physical sensations is compromised.

4. **Low insula activation may contribute to dissociation.** In addition to producing alexithymia, lowered insula activation may lead to dissociation, where you lose a sense of self-awareness and presence, "space out," or experience a time lapse where you feel as though you've disappeared for a brief period. Sometimes dissociation is also associated with feelings of being unreal, or feeling as though things around you are surreal. You may feel disconnected from yourself and your surroundings when this occurs. Alexithymia, numbing, and dissociation can make it difficult to connect with and understand yourself, and it can negatively impact your relationships, as you may appear distant and disconnected from those you love.

Use the following exercise to reflect on your experiences with alexithymia, numbing, and dissociation. You can also visit http://www.newharbinger.com/48848 to download a copy of this exercise.

Your Alexithymia, Numbing, and Dissociation Experiences

Insula deactivation is a common and understandable reaction to trauma, and it's often not within the survivor's control. Each person's experience of insula underactivation is a bit different. Answer the following questions to better understand how your own insula underactivation may be apparent, and how it might be impacting your life.

Do you find that you feel cut off from other people or your own emotions (or yourself more broadly) much of the time? What is that like? When do you tend to feel more or less cut off from yourself or others? For instance, some survivors notice that they feel numb when they become stressed or when people try to get too close to them.

_____

_____

_____

_____

Is it difficult for you to name, describe, or discuss your emotions? Was it always like that in your life? Is this something you avoid? If so, why?

_____

_____

_____

_____

_____

Do you sometimes feel as though things around you are surreal or that you're not real? Or do you at times feel as though you "space out" or disappear for periods of time? For example, many people have had the experience of getting in the car and driving, and then suddenly arriving at their destination without remembering how they got there. Do you have frequent experiences like this? If so, when do they tend to occur?

_____

_____

_____

_____

_____

Terms Your Therapist Might Use to Explain This Challenge

If your experiences are similar to Claire's, a therapist may use the following descriptors and terms in your discussions, or in your therapy notes:

- **Numbing symptoms.** Numbing symptoms are considered PTSD symptoms and are reflected in the *DSM-5* PTSD symptom cluster called "negative alterations in cognition and mood." If you experience numbing symptoms, you may find that you have difficulty feeling both pleasant and aversive emotions, and that you feel disconnected from others, even people you care about deeply.

- **Alexithymia.** This is considered a frequent phenomenon experienced in multiple disorders, but it is not a disorder or specific symptom itself. Alexithymia occurs when you have a hard time feeling, expressing, naming, describing, or expressing emotion.

- **Dissociation.** This is a common phenomenon after trauma that occurs when you feel disconnected from yourself, others, or your surroundings. Some people experience this as a zoning out or a sense of temporarily disappearing from the present moment. It can occur anytime, but it is commonly experienced when something reminds you of the traumatic event. Dissociation is considered a PTSD symptom, according to the *DSM-5*.

What Your Brain Needs

As is true with other brain regions, the good news about the insula is that it can be altered (structurally and/or functionally) so that you can begin to experience emotion and better connect with yourself, others, and the present moment. This can be accomplished through the consistent practice of various psychotherapeutic and mindfulness techniques. If you find your sense of apathy, alexithymia, or dissociation to be distressing, here is what your brain needs to feel more connected, self-aware, and present.

1. Bottom-up techniques that increase activation of the interoception center of the brain (the insula). As described earlier in this chapter, the insula is involved in helping you feel your feelings, through increasing awareness of physical sensations. When you're able to increase insula activation, experiencing, becoming aware of, and describing emotions becomes easier, and those are the goals of the bottom-up techniques described in this chapter.

2. Top-down techniques that help you focus on your interoceptive experiences. A great way to improve your ability to feel your feelings, and strengthen the insula, is to intentionally concentrate on, and maintain focus on, different sensory experiences or regions of the body. This focus requires prefrontal cortex activation, which can strengthen the prefrontal cortex while also connecting you more deeply with your body, improving insula activation.

The next section details five skills that contain strong bottom-up components that can help you learn how to activate your brain's "interoception center" (insula).

Skills for Meeting the Challenge

In this section, you'll learn five skills that can help you activate the insula, feel more present, and experience a deeper connection with yourself and others. These skills include grounding techniques, ways to connect with the body, and an exercise that improves emotional awareness by helping you access past emotional experiences and memories.

Skill 1: Outside-In Interoception (Bottom-Up)

Interoception is the ability to feel into and notice internal experiences such as physical sensations. When insula activation is low, interoception becomes difficult, which in turn makes it hard to recognize and describe emotions. In this exercise, you'll be given several grounding tips for connecting with your external environment to jump-start your insula. Specifically, this exercise provides several quick and easy ways to engage in strong (but safe) sensory experiences that will immediately awaken your insula and reconnect you with yourself and your environment. This is particularly helpful during times when you feel spaced out, disconnected from the present moment, or dissociative.

In therapy, it is common to practice grounding techniques to reduce dissociative experiences. These grounding techniques are sensory awareness exercises where you activate a sense (touch, vision, etc.) by engaging with something in the external environment. For example, you may be asked to identify five purple items in a room. In this instance, you activate sight and focus on specific types of objects in your environment. As you focus on your experience of what you see, hear, etc., your insula activates. This is because in connecting with your external environment, you are also connecting with your body as you attend to your experiences of things you are sensing. This type of practice can be extremely helpful when you want to feel more grounded, aware, and connected to yourself, others, and your surroundings. Here are some grounding ideas that, due to their sensory intensity, are particularly effective when you're feeling dissociative.

1. **Smell something very strong.** Not all sensory experiences are created alike, and not all sensory experiences, or grounding techniques, will be effective in reducing dissociation. The most effective grounding techniques are the ones that produce strong sensory sensations, and smelling something strong is arguably the best way to quickly activate the insula. This is because smell, as a sense, is sent directly to the amygdala and is not first filtered through other brain regions. This means that smell is very emotional and powerful as a sense. Thus, one way to effectively ground yourself is to smell something strong. Here are some ideas for strong smells you can use for grounding:

   - Essential oils
   - Garlic
   - Peppermint/mint
   - Citrus
   - Hand sanitizer
   - Strong perfume

2. **Taste something very strong.** Just as scent can be a strong sensory experience, so can taste. Here are some ideas for things to taste, which can produce strong sensory experiences and be helpful for grounding:

- A strong breath mint or mint gum

- Citrus

- Sour candy

- Spicy foods

- Flavorful hard candies

- Foods with cloves, chile, cinnamon, garlic, or ginger

3. **Create a strong tactile experience.** Touch can also be used to create intense sensory experiences. This is one reason why it is common for trauma survivors to self-harm, where they burn or cut themselves—these tactile experiences are very strong and can reduce dissociation and make the person feel more present. However, you want to be able to stay present and connected *without* self-harm! Here are some ideas for safe but strong tactile experiences. You'll notice each of them involves temperature, because temperature changes are a great way to produce strong tactile experiences without harm.

- Hold a piece of ice in your hand(s) until it completely melts, or hold ice packs in your hand(s)

- Splash your face with cold water

- Stick your face into a freezer for a moment

- Take a warm or cool bath

- Place a warm pad or blanket on your stomach

- Place one hand in cold water and the other in warm water, and notice the difference between the two

- Hold a hand warmer in your hands

- Step outside if the temperature is very different from your indoor temperature

TIPS FOR OUTSIDE-IN INTEROCEPTION

While the other skills in this chapter can be practiced on a regular basis, perhaps even daily, these interoception exercises are best used as needed, during times of dissociation or disconnection. The main reason is because they are intended to kick-start the insula in order to bring you back into your body and the present moment, but they are not meant to build the insula long term or increase baseline insula activation. Also, practicing these techniques daily, for long periods of time, may be very uncomfortable and cause distress (if, for instance, you were holding ice in your hands for thirty minutes straight). Thus, you should use these techniques during moments when you feel zoned out and you want to jolt yourself back to the present moment, and back into your body.

Skill 2: Body Awareness (Bottom-Up)

One of the best ways to activate and strengthen the insula is to feel directly into the body, noticing sensations in different regions. This technique, which is categorized as a type of body scan, will help you learn how to connect with different areas of your body and begin noticing how each region feels different from other areas. Additionally, in this technique you will learn how to engage your senses while noticing experiences occurring in different parts of your body.

1. **Prepare for the exercise.** Set up the space around you so that you have a few different items nearby that you can touch, smell, taste, see, and hear. For each sense that you intend to engage, have two items you can interact with. For instance, if you plan to use your hearing and sense of smell, have two items you can hear and two different items you can smell. This is helpful because you can compare and contrast the two. Examples of these items include lavender and a chocolate bar, for smell; a soft blanket and a cold table, for touch; the outdoors when you look one direction out of a window and a bedroom when you look in the other direction, for sight; a piece of popcorn and a piece of hard candy, for taste; and music on your phone and a TV show, for sound. Choose at least one sense to engage with during this practice, and prepare any items you would like to use.

2. **Scan your body.** Assume a comfortable position and close your eyes. Direct your attention to your feet and begin to feel into this region with openness and curiosity. As you feel into your feet, ask yourself the following questions:

- What sensations in this region do I notice?

- What other experiences do I notice in this region?

- Does this region feel warm or cold?

- Does this region feel heavy or light?

- Is there pain in this region?

- If this region had a texture, what would it be?

- If this region had a color, what would it be?

- If this region had an emotion, what would it be?

- What is it like to connect with this region?

Next, gently move your attention upward, to your lower legs, and repeat the above questions. You may repeat these questions for each region of your body as you slowly move your attention more and more upward. Here is a list of regions you may wish to feel into and connect with:

- Feet
- Lower legs
- Upper legs
- Hips/buttocks
- Hands
- Arms
- Lower back

- Abdominal area
- Chest
- Upper back
- Shoulders
- Neck
- Face/head

3. **Feel into your entire body.** Now that you have connected with, and scanned, these separate regions of your body, take a moment to notice any sensations or experiences that might be occurring anywhere in your body. Instead of trying to focus on just one area of your body, let your awareness drift to any region that calls your attention. Sit for a few moments just noticing anything in your body that is being communicated to you.

4. **Incorporate a sensory experience.** Next, identify a sense that you would like to connect with. This can be sight, taste, hearing, smell, or touch. Locate the items you prepared for this exercise and begin to engage with one of them. For instance, if you want to engage with your sense of smell, smell one of the objects you prepared (such as a chocolate bar or flower). Take a moment to sense this object, smelling it, holding it, looking at it, listening to it, or tasting it. As you engage with the object, pause to fully savor it. You may ask yourself the following questions about this experience:

- What is it like to [smell/feel/taste/hear/touch] this?

- Where in my body do I notice the experience of [smelling/feeling/tasting/hearing/touching] this?

- How does this [smell/feel/taste/sound/touch] make me feel?

- What emotions does this [smell/feel/taste/sound/touch] bring up in me?

- Are there any memories that I associate with this [smell/feel/taste/sound/touch]?

5. **Incorporate a second sensory experience.** Repeat step 4 with a second sensory experience. This second sensory experience should use the same sense as before (a second object to smell, touch, etc.). As you connect with this second object, you may ask yourself the questions listed in step 4 in addition to the following questions:

- Does this [smell/feel/taste/sound/touch] feel different from the last object? If so, how?

- How do I feel about this object, in comparison to the last?

- Are there different emotions or memories that this second object brings up for me, as compared to the first?

- How is this object similar to, or different from, the last one?

6. **Repeat.** Repeat steps 4 and 5 with another sense, if you wish. For example, if you began with smell, you might repeat these steps with touch. Repeat this process as desired.

TIPS FOR BODY AWARENESS

Body awareness is an exercise that can be practiced very briefly, over a five-minute period, or for a longer amount of time, up to thirty minutes. It is best to practice this technique, or components of this technique, on a daily basis when your goal is to increase baseline insula activation. If you'd like, you may alternate this technique with other techniques in this chapter. Also, if you find it difficult to answer some of the questions posed in this technique (such as asking about emotions that accompany a sensation), that is okay, and even expected. With time and practice, you may find that your ability to answer these questions improves as your connection to your body deepens. As this happens, and you can begin to notice, name, and describe experiences in your body, your emotional awareness will become stronger!

Skill 3: Heart Connection (Bottom-Up)

This technique, which is an adaptation from Sweeton (2019), teaches you how to connect with an internal function of the body, the heartbeat, by connecting with the outside of the body. It is a great way to more deeply connect with your body and increase insula activation.

1. **Begin to breathe deeply.** Close your eyes, if you feel comfortable, and begin to bring you awareness inward, noticing what it feels like to inhale and exhale.

2. **Feel into your heart.** As you continue to breathe, shift your attention to your heart inside your body, just noticing any sensations you may be experiencing in this region.

3. **Find your pulse.** To connect more deeply with the heart, take the fingers of your right hand and press gently on the inside of your left wrist, locating your pulse. When you feel the pulse, begin to imagine the beating that you feel as occurring both in your wrist, where you can feel it with your fingers, and in your heart region. Continue connecting with your pulse for a few moments, imagining your pulse to be located in your wrist and your heart.

4. **"See" your heart in your mind.** As you continue to notice your pulse, you may also bring into your mind's eye an image of your heart. The image may be of your heart inside or outside your body. "See" your heart in your mind's eye, and as you look at

it, imagine that it is beating in tandem with your pulse. Stay here for a few moments. When your mind wanders, gently return your focus to the thumping of the pulse in your wrist and the image of your heart in your mind.

TIPS FOR HEART CONNECTION

You can think of this technique as a sort of meditation and breathing exercise that incorporates body awareness and body focus. Techniques of this sort are great for activating the insula. However, to strengthen the insula, repetition, time, and focus are required. You should practice this exercise, or other insula-building exercises, daily for ten to fifteen minutes total per day. This exercise, and others in this chapter, can be broken down into five-minute or even three-minute increments, for ease of practice.

Skill 4: Physical Profiles of Emotions (Bottom-Up and Top-Down)

This advanced insula-building technique has you create physical profiles for different emotions, which are induced through memory recall. You may have noticed that sometimes when you think of past emotional experiences, you can feel those emotions all over again. Trauma survivors frequently experience this with regard to traumatic memories, but you can also reexperience emotions attached to other types of memories, such as pleasant ones. In this exercise, you will intentionally bring to mind multiple emotional memories, each of which elicits different feelings. Then, you'll map out in the body where you experience those emotions most strongly, thus creating a sort of physical profile for that specific emotion. For example, you may notice that when you think about a memory where you missed out on a promotion at work, you experience anger. This anger may be associated with chest tightening, rapid heart rate, fists clenching, and the stomach feeling hot. In that case, your physical profile of anger may look like the following example.

Physical Profile of Emotions Example

My Physical Profile of: ___Anger___

Rapid heart rate _____

Fists clenching _____

Chest tightening _____

Stomach feels hot _____

_____

_____

_____

_____

Use the worksheet following the directions to map your own physical profile of emotions. You can also visit http://www.newharbinger.com/48848 to download a copy of this exercise so you can complete it with multiple emotions.

1. **Identify an emotional memory.** Bring to mind a memory of something that was emotional for you. This can be a memory that elicits one of the following emotions:

Happiness	Contentment
Excitement/joy	Love
Surprise	Anger
Sadness	Fear

   If possible, select a memory that brings up one emotion in particular, as opposed to multiple emotions. This will ensure that when you create your physical profile, it will reflect the physical experiences of *one* emotion, as opposed to several.

2. **Replay the emotional memory.** Begin to reconnect with the memory, replaying it in your mind from start to finish. Close your eyes as you do this, if it is comfortable. To fully connect with the memory, try to recall not only the facts of the memory (who, what, where, when) but also the sensory experiences you had at the time (what you saw, heard, felt, etc.). Stay here for a moment, fully accessing and replaying this memory.

3. **Notice your emotions.** As you keep the memory in mind, see if you can feel, even just a little, any emotional reactions to it. Does it still elicit anger? Joy? Connect with this emotional reaction to the memory.

4. **Notice your body.** Continuing to connect with the emotion tied to the memory, shift your attention to your body, noticing any sensation or experience occurring as a result of this emotion. You may notice, for instance, that your breathing or your heart rate has changed. You may also notice temperature changes (feeling hot/cold) or pressure changes (feeling heavy/light). If you'd like, you may scan your entire body, becoming aware of how different regions of the body react when this emotion

is felt. Here are some common experiences you can check in with yourself about as you scan each area:

a. **Temperature changes:** Different areas feeling hotter or colder.

b. **Pressure changes:** Different areas feeling heavy, light, or compressed.

c. **Texture changes:** Different areas feeling soft, sharp, or rough.

d. **Movement changes:** Different areas feeling as though energy is moving them up or down (for instance, the sense that the stomach is dropping).

5. **Map out the sensations.** Map out the physical sensations that you notice as you access this emotional memory. Write down the sensations on the right side of the body outline and draw arrows to the parts of the body you experience those sensations in.

Physical Profile of Emotions

My Physical Profile of: _____

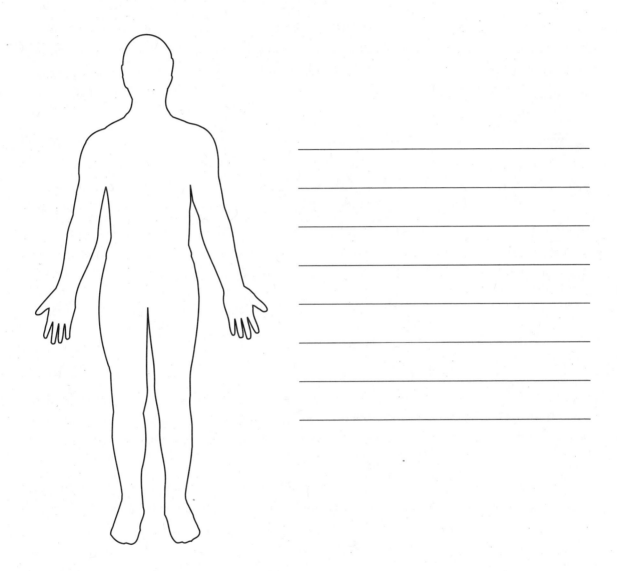

_____

_____

_____

_____

_____

_____

_____

TIPS FOR PHYSICAL PROFILES OF EMOTIONS

This exercise can be repeated for multiple types of emotions. Complete this technique for happiness, love, sadness, anger, and fear. When you are able to establish a connection between your emotions and sensations, you can better connect with, name, and describe emotions, and it reduces alexithymia and dissociation. After creating your physical profiles of emotions, refer back to these profiles when you notice emotions arising, so that you may discern additional sensations over time. Once you become aware of how emotions feel in your body, you can recognize emotions more easily.

When you experience distressing emotions, you may wish to disengage a bit from some of those sensations. To do this, however, you first have to know where distress is located. When you notice it, you may briefly check in with the distress, and then move your attention to other areas of the body that you've associated with calmness or more pleasant feelings. For example, you may notice a heaviness in your stomach when you're sad, and you may check in with this sensation briefly before moving your focus to thoughts of someone you love and the accompanying warmth you feel in your chest when you access those loving feelings.

Skill 5: Practicing Feeling (Top-Down)

Have you ever watched a movie and felt sad? Or heard a song and felt energized? If so, you know how emotion can be accessed and intensified through music, TV shows, movies, plays, other forms of entertainment, and even other people. However, when feeling numb, apathetic, or upset, it can be hard to remember what to watch or listen to, or whom to talk to, to feel better.

A top-down strategy for connecting with emotions is to intentionally induce those emotions by focusing on watching movies, listening to music, or engaging in activities that have historically prompted strong emotions. You can think of this as a way to practice feeling with the help of prompts, situations, entertainment, or people that usually get you emotionally activated in some way. Of course, you want to be careful about purposefully inducing distressing emotions, as that is not the goal here! Rather, the goal is to identify the activities that tend to elicit a sense of joy, humor, calm, or contentment.

Use the following worksheet to help you identify ways to "practice feeling," so that you can refer to it when you want to intentionally access strong, positive emotions. You can also visit http://www.newharbinger.com/48848 to download a copy of this exercise so you can do it with multiple feelings.

Practicing Feeling Worksheet

Answer the following questions for the experience of *love*:

Do you have pictures that activate feelings of love when you look at them? If so, where are these pictures? (This may include pictures of babies, children, pets, or a partner.)

_____

_____

Is there a type of music, or a particular song, that elicits feelings of love? What is it? (This may be a song you danced to at your wedding or a song that reminds you of someone you love.)

_____

_____

_____

Are there movies or TV shows that remind you of what love feels like or that remind you of someone you love? What are they?

_____

_____

_____

Are there smells, such as the smell of baking or a candle, that elicit feelings of love? What are these smells?

_____

_____

_____

Are there physical items that, when you touch them or see them, remind you of someone you love? This can be a way of connecting to others. What are those items?

_____

_____

_____

Are there people you can talk to, see, or call who elicit feelings of love or make you feel loved? Who are they?

_____

_____

_____

Repeat each of these questions for joy, peace, contentment, humor, and amusement and record your answers on a piece of paper or on a notes app on your phone, for easy access. When you notice that it is difficult for you to access positive emotions, you can refer to these notes as a way of jump-starting your emotions!

TIPS FOR PRACTICING FEELING

Try to generate ideas for practicing feeling when you're relatively calm and clearheaded. This way, you're more likely to be able to come up with some great ideas. If you wait until you're distressed to give this thought, it may be difficult to come up with any good ideas, as you might be too overwhelmed. Also, revisiting this worksheet regularly can be helpful, as new ideas might come to mind.

A Pause to Reflect

Now that you've come to the end of this chapter, it is a good time to pause and reflect for a bit on what you've learned and the skills that you've found helpful. Note what works and what you may want to keep practicing moving forward to establish a consistent practice of the skills that will strengthen your insula. In the next section, you can list the skills that you've found helpful. In this list you may include skills you've learned in this chapter as well as any other body-based or body-focused exercises you enjoy, as any grounding or body-based technique is likely to increase insula activation. You can also visit http://www.newharbinger.com/48848 to download a copy of this exercise.

Skills for Reducing Apathy, Alexithymia, and Dissociation

To feel more present and connected, I can try one or more of the following, which have been helpful to me in the past.

Skill/Technique	Best Time to Practice	When I Practiced the Skill (Day and Time as Well as Description of the Situation)	Reflect on the Experience. Was the Skill Useful?
Outside-In Interoception	When feeling numb or apathetic or as a regular practice.		
Body Awareness	When feeling numb or apathetic or as a regular practice.		
Heart Connection	When feeling anxious or numb or as a regular practice.		
Physical Profiles of Emotions	Complete once and then again every few months as desired, to understand how emotions are represented in the body.		
Practicing Feeling	When feeling numb or apathetic.		

CHAPTER 6

Issues Around Safety and Control

Hailey is an overachieving twenty-seven-year-old graduate student who came to therapy after being raped in college about five years ago. Even though Hailey lives in a safe part of town with trustworthy roommates and her supportive parents are nearby, she can't seem to ever feel safe. Over time, she's become increasingly isolated as a result of feeling out of control, and she said she tends to lose her sense of control when she leaves her apartment. Hailey has said that it seems like the assault robbed her of her peace of mind, and she's afraid she'll never be able to feel safe again.

Check-In: Your Experiences

After reading about Hailey, take a moment to think about your experiences. You may choose to reflect and journal on the following questions:

- Have you ever felt similar to Hailey, as though you can't seem to feel safe? Or perhaps you only feel safe when you feel totally in control?

- Do you notice that you feel anxious if you're not in control of a particular situation, yourself, or others? How do you think trauma might be contributing to this?

- Do you ever worry that you've been robbed of your sense of safety, or perhaps robbed of something else, as a result of your trauma(s)? If so, what?

- Do you believe you have to remain in control in order to feel safe?

How Safety and Control Issues May Impact Your Life

Themes of control and safety are almost universal in the lives of trauma survivors. And right-fully so: when you are in complete control of something, you can ensure you stay safe, right? It makes sense to prioritize safety above all else when you've experienced the consequences of being *unsafe*, especially when bad things happened as a result of other people taking away your control (and the ability to protect yourself!). Thus, safety and control are common and inter-related themes after trauma. You rarely encounter one of these themes without the other.

While there is nothing unhealthy about desiring and pursuing a sense of safety and control, these efforts can become unhelpful, distressing, or disproportionate for some people. For example, it is common for trauma survivors to fear going certain places or engaging in certain activities, even when they know those things are relatively safe. This can rob them of the ability to live the "normal" life they previously enjoyed. After trauma, your safety "alarm bells" may go off even when you aren't unsafe, creating anxiety about living your life in ways that previously felt comfortable. While it used to be that you could go out with friends, or drive at night, or travel, or have close relationships, after trauma these things can become difficult if they begin to feel unsafe to you.

In response to feeling unsafe, many people attempt to exert control in their lives as a way of reestablishing safety and lowering anxiety. This makes logical sense, as we all want to be proac-tive about staying safe! However, when you exert control in unhelpful or ineffective ways, it can cause problems at work or in relationships. Those around you may complain that you are too rigid with your plans, or that you try too hard to control the relationship or the other person's behavior. You may also have difficulty delegating tasks to others or letting go of "gatekeeping" tendencies in a work context. These tendencies can cause additional stress for you and those around you.

The following self-assessment is intended to help you become more aware of the roles that control and safety play in your life and to determine whether these themes have become prob-lematic for you. You can also visit http://www.newharbinger.com/48848 to download a copy of this exercise.

Safety and Control Issues Self-Assessment

Read the following and circle the best response from 0 to 3, where 0 = None/never, 1 = A little bit/ sometimes, 2 = A moderate amount/often, and 3 = A lot/most of the time.

I feel anxious when I can't control social plans.	0	1	2	3
People have told me that I'm a "control freak."	0	1	2	3
I have a hard time delegating responsibilities to others, even when I know it is their job or that they are able to do the tasks.	0	1	2	3
When I can control a situation, I feel a strong sense of relief.	0	1	2	3
I avoid situations that seem unstructured or unpredictable.	0	1	2	3
I feel anxious when I cannot "account" for everything happening in a situation (for example, when I cannot see everyone in a crowded room or know everything happening at a particular moment).	0	1	2	3
I exert control in situations, or with people, to feel safe.	0	1	2	3
When I feel unsafe, I'll do anything to reestablish a sense of control.	0	1	2	3
I try to control things even if there's no need for me to.	0	1	2	3
People get frustrated with me because I have to have things done the "right way."	0	1	2	3
I do things myself because if I delegate the task, others will do it incorrectly.	0	1	2	3
I have a hard time "going with the flow."	0	1	2	3
I am the primary decision maker, or leader, in my close relationships.	0	1	2	3
I feel as though something bad is about to happen even if there is little evidence for this feeling.	0	1	2	3
I can't feel safe.	0	1	2	3
The only person I can trust to keep me safe is me.	0	1	2	3
I worry about being unsafe in a lot of contexts.	0	1	2	3
I don't do things I used to like to do because I worry about being safe.	0	1	2	3
I spend a lot of time trying to ensure that I will stay safe.	0	1	2	3

If I don't protect myself, I'll get hurt by others.	0	1	2	3
I am better than most people at seeing what could go wrong in a situation.	0	1	2	3
If I'm not vigilant, bad things will happen.	0	1	2	3
If something bad were to happen to me, it would be because I didn't do enough to keep myself protected.	0	1	2	3
I've been told that I am overprotective of the people I love.	0	1	2	3
I am a "helicopter" parent or partner.	0	1	2	3
I avoid social or work-related events because I am anxious about bad things happening.	0	1	2	3
If I can anticipate what might go wrong, I can stop it from happening.	0	1	2	3
I'll stay safe as long as I'm in control.	0	1	2	3
Total				

Add up the total score from the above items, and record it next to "Total." A score of 33 or higher indicates that themes of safety and control might be causing you problems in your life.

Use the following exercise to reflect on your self-assessment score. You can also visit http://www.newharbinger.com/48848 to download a copy of this exercise.

Reflect on Your Self-Assessment Score

What did you score on this self-assessment? Were these things you already knew about yourself? Consider the following questions to reflect on what you just learned about yourself.

Do you see a connection between control and safety? Had you considered this before?

_____

_____

Were your scores different from what you expected? Did you learn anything new from the exercise?

_____

_____

_____

Which theme plays out the most in your life: control or safety? How do you see this topic impacting your life?

_____

_____

_____

Are there shifts you would like to make with regard to safety and control in your life? Are there fears you have about making these changes?

_____

_____

_____

Why Issues of Safety and Control After Trauma Are So Common

As noted in other chapters, your brain is designed to keep you alive, *not* calm, happy, or even mentally well! When your safety has previously been threatened, or something terrible has happened where you've been unsafe, the brain will work extra hard to ensure you stay safe in the future. At times, this can feel extreme, like the brain is prioritizing safety and survival at all costs. And it probably is. When you experience awful consequences resulting from being *unsafe*, your brain becomes very good at recognizing future situations where danger might be present. These situations need not be exact replicas of the past traumatic event(s); rather, even a small common element or experience may lead your brain's alarm bells to sound. For instance, if you've experienced a car accident, you may not only develop a fear of driving or of being a passenger in a car, but also become afraid of any situation where you're not able to control others' behaviors or of any context where you feel trapped, leading to claustrophobia.

In this chapter's example, with Hailey, after being raped her brain's danger alarms weren't contained to situations where she might be assaulted again. Rather, her brain started sounding alarms even when she considered leaving her apartment. While going to the grocery store may not overtly resemble rape, and may not even reflect an explicit fear of being raped again, Hailey's fear about not being in control extended to any situation where she may feel out of control. Going to the grocery store is a relatively safe behavior for most of us. But for a rape survivor, any public outing, where other people may behave unpredictably, can feel terrifying. Thus, Hailey may find herself avoiding a lot of different situations in order to feel safe and in control of her surroundings.

The brain's response to trauma drives these fears and "triggers" about safety. After trauma, the amygdala, the threat detection center of the brain, becomes sensitive to anything remotely resembling the original trauma(s) and "sounds the alarms" whenever it detects something that looks or even feels similar to traumatic events of the past. A common response to these triggers is to exert control in the situation, because a perceived sense of control can quiet the amygdala, making you feel less fearful, agitated, or triggered. In other words, your attempt to control your life and the situations you enter is a great way to (temporarily) calm your amygdala and reestablish a sense of safety!

As noted earlier in this chapter, that's not necessarily an unhealthy thing. Choosing, for instance, to take a cab rather than walk alone in a dangerous part of town is a way of taking control of a situation to keep yourself safe. But when efforts to control situations are taken too far, they can negatively impact your life. Also, a lot of trauma survivors find that while a sense of control can make them feel better in the moment, it often doesn't solve the larger problem of feeling unsafe in general; over time, their control efforts become less and less effective and they become increasingly less tolerant of ambiguity or uncertain situations. This, in turn, may lead them to exert even more control, in more extreme ways, but these efforts also usually fail to establish a long-term sense of safety. To better understand how this cycle of maladaptive control efforts works, here's a summary of what many trauma survivors experience.

1. **You encounter a trauma reminder or think about a situation that might be unsafe, and the amygdala activates.** This amygdala activation alerts you that what

you're considering doing (going to the grocery store, to a party, etc.) might pose a danger that will threaten your safety. When a trauma survivor is "triggered," they are usually experiencing amygdala activation and the cascade of reactions this activation creates, including the body going into the stress response. The brain then very quickly weighs its options for how to handle the situation, and may choose to "fight," "flee," or "freeze" in response to the danger. Not surprisingly, if this is a situation that has not occurred, but is one that you're merely considering (such as going to a sporting event next week), your brain will recommend the "flight" response, where it encourages you to avoid it in order to stay safe. However, your brain may also consider ways for you to be able to attend the event, if you can somehow convince your brain that you'll be safe in doing so.

2. **To convince your brain to allow you to enter the situation, you engage your prefrontal cortex to consider ways you might stay safe.** Amygdala activation can make the "logical brain" go offline, making it difficult to reason through different scenarios. However, when amygdala activation is anticipatory, meaning that it's in response to a future theoretical event, there is often some room for neural negotiation, where you can engage your prefrontal cortex in an attempt to quiet your amygdala. This occurs when you try to convince yourself that a situation will be okay, or that something isn't that bad, or any other mental effort to make yourself feel better. It's as if your prefrontal cortex is trying to reason with your amygdala in order to downregulate it and make you feel safe!

3. **Specific control-type behaviors or safety measures are identified.** A common strategy to accomplish this amygdala downregulation is to try to convince the amygdala that the situation will be safe because you can somehow control it. Who better to control the situation than you? Would that not ensure that you stay safe? Thus, it's often the case that trauma survivors find themselves strategizing about safety measures they can take and ways they can influence or control the situation to make sure nothing bad happens. For example, you may eat out at restaurants as long as you can see the door or sit with your back to the wall. Or you may only go to a movie when you can sit next to an emergency exit. In social situations, you may only attend if you can control the time, place, and attendees, and may refuse to attend if plans change or are "up in the air."

4. **The amygdala temporarily deactivates.** When you're able to convince the amygdala that you'll be able to stay safe, it will likely calm down and you'll feel a bit better. However, even if you can convince yourself that you can control the situation, the relief you feel will usually be temporary and will be followed by amygdala reactivation.

5. **When confronted with the situation, the amygdala reactivates.** This reactivation, signaling potential danger, often occurs despite your safety precautions or exertions of control. While it's possible that your sense of control will allow you to enter the situation, it is unlikely that you will feel at ease in it or that the amygdala downregulation will last very long. Most survivors experience a sort of rebound effect, where the fear comes back, in a jolting manner, regardless of the steps they've taken to stay safe. The result may be that you take even more extreme steps to stay safe, including exerting more control over the situation. However, the effectiveness of these attempts, too, will often be short lived.

The ways in which you attempt to stay safe can become habits, much like expressions of anger or coping strategies. However, it can be difficult to consciously recognize these behaviors and determine when they need to be shifted. Use the following exercise to reflect on how you exert control to stay safe. You can also visit http://www.newharbinger.com/48848 to download a copy of this exercise.

Your Safety and Control Behaviors

Consider the following questions as you reflect on ways you might exert control in order to establish, or reestablish, a sense of safety.

When your "alarm bells" go off in your brain about a particular situation being unsafe, what steps do you take to make it seem more safe? Do these behaviors take a lot of time, energy, or resources?

_____

_____

_____

Do any of those steps or safety behaviors involve an attempt to increase your sense of control? If so, are they usually successful? Do they often cause problems for you?

_____

_____

_____

Over time, do your safety behaviors tend to lead you to feeling more safe? Or do you seem to return to feeling unsafe, regardless of the steps you take?

_____

_____

_____

Do you find yourself becoming more extreme in your safety or control behaviors over time? If so, how? What is the result of this?

_____

_____

_____

It is possible that you may be engaging in safety and control behaviors that work well for you and accomplish what you need them to. If so, that is great! However, if you find that these behaviors take a lot of time, energy, or resources, or that they are ineffective or otherwise problematic, it might be time to find other strategies to feel safe.

Terms Your Therapist Might Use to Explain This Challenge

If your experiences are similar to Hailey's, a therapist may use the following descriptors and terms in your discussions, or in your therapy notes:

- **Negative alterations in cognition and mood symptoms.** This is one of the symptom clusters of PTSD according to the *DSM-5* and it includes seven symptoms, each of which describes changes in how a trauma survivor thinks and feels. For example, you might feel isolated or notice difficulty accessing positive emotions such as love or happiness, which are emotional changes that are common after trauma. Additionally, you may experience changes in your thinking, such as becoming more suspicious of others or having more extreme negative thoughts or expectations. You might, for instance, perceive relatively safe situations as dangerous.

- **Arousal and reactivity symptoms.** This, too, is one of the symptom clusters of PTSD, according to the *DSM-5*. The six symptoms of this cluster include two that are of particular relevance here: hypervigilance and sensitive startle response. Hypervigilance occurs when your brain becomes overly on guard for danger and prone to detecting danger in benign situations. A sensitive startle response is similar; it occurs when you feel suddenly shaken, or startled, in response to unexpected, small, usually innocuous noises or surprises.

What Your Brain Needs

After trauma, your brain's danger "alarm bells" may easily sound, making you feel unsafe even in relatively safe situations. This can lead to control behaviors that actually work against you. While they may make you feel safer in the moment, they ultimately increase fear and anxiety because they don't allow you to learn (or relearn) that these situations are safe. Here is what your brain needs for you to be able to establish a reasonable sense of safety and control.

1. Bottom-up techniques that reduce the activation of the amygdala, which is the "threat detection center" of the brain. Just as dysregulated anger is often a result of the amygdala misinterpreting situations as being more dangerous than they really are, feeling unsafe in relatively safe situations also reflects an amygdala error, where it incorrectly

concludes that something, or someone, is threatening. This is common after trauma, which is associated with overactivation and hyperreactivity of this brain structure. To help regulate and soothe the amygdala, you can practice bottom-up techniques.

2. Top-down techniques that activate the "thinking center" of the brain (prefrontal cortex) and the "memory storage unit" of the brain (hippocampus). Although thinking of ways to manage or control a situation, which involves the brain's "thinking center," can be effective, sometimes it can work against you. However, that doesn't mean that cognitive techniques—skills that use your prefrontal cortex and require intentional thought and effort—are unhelpful. In fact, some mental exercises can be quite effective, it just depends on the strategy you're employing!

The next section details four skills that can help you manage feelings of being unsafe and help you take control in effective, reasonable ways.

Skills for Meeting the Challenge

This section focuses on safety- and control-themed skills that can help you downregulate the amygdala (the threat detection center) using your prefrontal cortex (thinking center) and hippocampus (memory storage unit). These skills include a body-focused exercise, resource tapping, and two cognitive techniques. Additionally, the Physical Profile of Emotions worksheet in chapter 5 can be a great tool for better understanding how emotions, and other feelings such as safety, are experienced in the body. You may use that skill to learn about the physical sensations associated with feeling safe and unsafe, and combine that with relaxation exercises such as deep breathing or autogenic training (described in chapter 1).

Skill 1: Accessing Control in the Body and Mind (Bottom-Up and Top-Down)

Just as you can create a physical profile of feelings of safety in the body (or a profile of what it is like to feel *unsafe*), you can also become aware of what it feels like to have a sense of control. This exercise helps you intentionally access the experience of control, noticing how it is represented in the body and experienced in the mind as thoughts. Learning how to intentionally access feelings of control, which often promote a sense of safety, power, and self-efficacy, can be a great way to bolster your strength and resource yourself before going into uncertain or anxiety-provoking situations where you may feel ineffective, weak, or unsafe.

1. Bring to mind a time in your life where you felt strong, in control, and confident. This may have been in an academic or a work setting, while practicing a sport, in a competition of some sort, or in a social situation. Bring this memory to mind, trying to picture it clearly. Imagine where you were, what you were doing, and who else was there. What were you seeing, hearing, and feeling? Take a moment to reflect on this memory.

2. As you hold on to that memory, begin to check in with your body, starting at your feet and slowly working your way up. What do you notice as you scan your body? This will tell you how this memory has been stored *in the body* and what a sense of control and confidence feels like in the body. Slowly move your attention from your feet to your lower legs, upper legs, core region, stomach, back, chest, arms, hands, fingers, shoulders, neck, and head. You may imagine this awareness as being a light beam moving upward through your body from the floor. What do you notice? Specifically…

   • Do you notice any warm or cool sensations? If so, where?

   • Do you notice any tension or relaxation in any regions? What is that like?

   • Do you notice any sense of pressure, heaviness, or lightness? If so, where?

   • Do you have a sense that these experiences have a shape or color to them? (Do not worry if this question does not resonate with you; that is okay!)

   • Do your sensations have a texture to them? Do any feel rough, soft, or smooth?

3. Now that you can connect with the physical profile of control, you may also reflect on thoughts or memories that may be associated with this experience. You may ask yourself the following…

   a. What words come to mind as you experience these sensations? Some people report noticing words such as "strength," "capable," or "calm." Does your brain associate any words with these physical sensations?

b.   What phrases or statements come to mind with these sensations? For instance, you may notice thoughts such as "I can do this," or "You got this," or "You're okay" coming to mind. Take a moment to check in with your brain about any statements tied to this physical profile of control.

c.   Does your mind bring to you any other memories when you access these experiences in your body? You may notice that when your body is reminded of what it feels like to be in control, you may access additional memories of times when you felt similarly. You may take a moment to jot down other times in your life when you felt these sensations and a sense of control.

TIPS FOR ACCESSING CONTROL IN THE MIND AND BODY

Now that you understand how a sense of control is represented in your body and mind, you can use this as a resource that you intentionally access when your confidence needs a boost or when you feel out of control or anxious. You may even practice this daily as a meditative exercise. Many people find that when they can access parts of themselves (through the body and/or mind) that feel solid, strong, and in control, it helps them face uncertainty with less fear.

Skill 2: Resource Tapping with Affirmations (Bottom-Up and Top-Down)

Bilateral movements, such as tapping, can be used to help you process distressing thoughts or memories. This is a common strategy used in eye movement desensitization and reprocessing (EMDR), where therapists have clients quickly move their eyes back and forth to facilitate processing and desensitizing to traumatic memories. However, most people don't realize that you can also use bilateral movement, such as tapping, for a different purpose: resourcing. In other words, bilateral movement can help you feel *less* distress about something upsetting or traumatic, but it can also help you feel *more* of something helpful or positive. It all depends on how the tapping, or movements, are done!

In particular, fast movements, such as eye movements or tapping, are frequently used to dull emotional experiences, making them less impactful. That's why bilateral movements are used in EMDR for trauma processing. However, *slow* bilateral movements can emphasize and magnify an experience, thought, memory, or emotion, making it more intense and impactful. This technique combines slow bilateral tapping with affirmations, or positive statements, to help you access and emphasize feelings of strength, confidence, control, and safety.

1. Bring to mind a memory of a time when you felt safe, secure, strong, confident, or in control. If there is another positive experience or memory that makes you feel grounded and happy, you may select that one instead. You may also consider using the same memory you brought to mind in the last exercise. What is the memory? What happened in the memory? What did you do and how did you feel? Spend a moment reflecting on this experience.

2. As you bring this memory to mind, close your eyes (if you choose) and gently begin to incorporate slow bilateral tapping, spacing the tapping by two or three seconds. You may tap each shoulder with your hands or rest your hands in your lap, alternating taps on the tops of your legs. Or you can tap your feet on the floor.

3. Bring to mind a positive, calming, or helpful statement, or affirmation, to repeat as you continue tapping and accessing the memory. Silently (or in a soft voice) and slowly repeat this affirmation for a few minutes. If you do not find it helpful to include the affirmation, you can let go of this piece for the exercise.

As you practice this technique, you may begin to feel the emotions linked to this memory more intensely. Some people also report feeling soothed by the slow, gentle bilateral movement, which can be rhythmic and calming.

TIPS FOR RESOURCE TAPPING WITH AFFIRMATIONS

Practice this technique when you want to more deeply access and experience positive emotions and memories, which can help bolster and strengthen you when you encounter stressful or uncertain situations. To experience the full effect of the exercise, practice it for five to ten minutes as a mindfulness practice. If it does not feel comfortable to do this with your eyes closed, you may practice with your eyes open, gazing at a spot on the floor.

Skill 3: How to Spot Safe People (Top-Down)

One common trigger for trauma survivors is relationships and interacting with other people. This is because a lot of trauma involves betrayal and harm that was inflicted on you by other people. Interpersonal harm, such as abuse, neglect, and betrayal, can rob you of your ability to feel safe and at ease around others. And rightfully so, as you'd never want to experience that pain again! In response to these traumas, a lot of survivors struggle when it comes to determining who is safe or unsafe. How do you know whether someone might be safe? What are the signals? Can you trust your judgment, especially if you've been wrong in the past? After trauma, it may feel like *no one* is safe, as your brain may send you danger signals even around trustworthy and safe people. This can make it difficult to tell what's true and what's a false alarm. Although it's impossible to tell with 100 percent certainty who is safe and who isn't, this short questionnaire can help you identify and examine some of the indicators for knowing whether someone is emotionally and physically safe. You can also visit http://www.newharbinger.com /48848 to download a copy of this exercise and do it with more than one person.

Are They Likely Safe?

When you want to do a quick "sanity check" about your assessment of a person, and whether they might be safe, ask yourself the following questions.

Do you know people who know this person? When you know other people who also know this person, it can be helpful, as others may be able to "vouch" for the person in question. If you don't know anyone who knows them, can you meet someone who does to get a sense of how they are viewed by the other person?

_____

_____

What have your experiences been like with this person so far? Have they done anything yet, that you can identify, to break your trust?

_____

_____

_____

How do you feel when you're around this person? When you're in their presence, what's it like? Although we can't always trust our "gut reactions," we often can, and these are important to consider. What sensations do you notice when you're around them?

_____

_____

How do you feel when you're *not* around this person? Is it similar to, or different from, how you feel when you're face-to-face with them? Sometimes we feel great when we are around someone, but then develop suspicions when we're not around them, after the fact. This can lead to your thoughts spinning out of control, but the thoughts may not actually be based on your real experiences with the person. Other times, however, you may find that you try to convince yourself to like the person, or feel safe with them, when you're *not* with them. Later, when you see them, however, you may notice that something just doesn't feel right. That, too, is important to consider.

_____

_____

What are your fears about this person? Do you have any evidence indicating that your fears may reflect something they've done or would be likely to do?

_____

_____

Does this person, as far as you know, have any history of violence, abuse, etc.? Do they have past charges, restraining orders, or legal problems?

_____

How does this person talk about other people? Do they speak about others in a disparaging manner? Do they seem to blame others when things go wrong?

_____

_____

Do you have any indications of how this person manages conflict? Have you seen them angry before, or know someone who can speak about how they handle anger or disappointment?

_____

_____

Does the person have long-term friendships or romantic relationships? Are there people this person is close to?

_____

_____

Have you ever said "no" to this person or attempted to set a mild boundary? If so, how did it go? Were you made to feel bad or uncomfortable?

_____

_____

_____

Does this person seem to be reliable? Do they show up when and where they agree to? More broadly, do their actions seem consistent with their words?

_____

_____

Is this person willing to trust _you_? Do you get the sense that they're open to being transparent and vulnerable with you?

_____

_____

TIPS FOR HOW TO SPOT SAFE PEOPLE

While this list is by no means exhaustive, it can give you a good starting point in assessing whether a person might be emotionally and physically safe. Review each of your answers in isolation, but also as a whole, because when you view your answers in totality, it can give you a different overall view of the person. This in turn can help your "rational brain" combine and consider these different factors to determine whether this is a safe person you want in your life. For example, it may be that you do not know anyone else who knows this person, which is not ideal. However, maybe you've noticed that this person is reliable, consistent, transparent, and vulnerable with you, and you have no reason to believe that they have ever been harmful to other people. You may have also set a boundary with them, and in response, they honored it. In addition, perhaps they frequently reference other people they are close to, and speak highly of others in general. Those factors, considered together, look a bit different than when you look at each in isolation.

Skill 4: Controlling What You Can (Top-Down)

Control isn't all bad! Sometimes you may get a message that if you try to control parts of a situation, you're a "control freak." However, efforts to control situations can be reasonable and preferable. Any parent will tell you that about managing children! It's up to you to determine what's "too much" control, but if you notice that others are complaining about your behavior or it's negatively impacting your relationships or ability to function as you'd like to, then you may need to reassess your efforts. The following exercise will help you explore what exactly you *can* control in a particular situation and how far you're willing to go—reasonably—to establish a sense of safety and control.

Use the following worksheet to evaluate what you can, and should, control. You can also visit http://www.newharbinger.com/48848 to download a copy of this exercise and complete it with multiple situations.

Assessing What You Can Control Worksheet

What is the situation that you want to exert some control over? What is supposed to happen? With whom, and when?

_____

_____

What are you concerned about happening in this situation? Specifically, what are you fearful about, or what about it seems like it might be unsafe? For example, maybe it takes place after dark or with people you don't know.

_____

_____

Looking at your answers to the above questions, what are the specific factors or details that make the situation potentially unsafe? Place a checkmark next to the details on the following list and use the extra lines to write in other factors.

- ☐ I am going alone.

- ☐ It's in a bad part of town.

- ☐ I don't know anyone there.

- ☐ Something bad has happened to me in a similar situation.

- ☐ I don't have a way to get home.

- ☐ I may not be able to get to an exit quickly if needed.

- ☐ _____

- ☐ _____

- ☐ _____

Referring to the previous list of factors, circle the factors you can *easily* either eliminate or mitigate, meaning there are ways of managing these factors without spending a large amount of time, effort, energy, or other resources. For each of these factors, write down how you plan to eliminate or mitigate them. For instance, if the situation includes all strangers, you may change this by bringing someone with you that you know (assuming that's an option for you).

Looking at the remaining factors (the ones you can't easily mitigate or eliminate), add one piece of information to each: What are the feared consequences of that factor? For example, if the factor is "If I go to the party, I'll be going alone, and I can't bring anyone else," the feared consequence might be "I might get physically assaulted at the party."

On a scale of 1 to 100, how unsafe do you feel when you look at each of these factors? You can assign an "unsafe score" to each of the factors you've identified.

You've already established that these factors can't be eliminated or mitigated. Referencing the example in step 5, perhaps you already know that if you go to the party you have to go alone, and there is no way to change this. And perhaps this makes you feel quite unsafe (an unsafe score of 50 on a scale of 1 to 100). However, when you look at the feared consequences, there may be ways to minimize or mitigate *those*. Using the same example, suppose that there is a party that is important for you to attend, and suppose that you will have to go alone. If you can't change those things, you can examine your feared consequences to see whether you can reduce the likelihood of *those* occurring. So, now the question becomes, "How do I prevent the terrible consequences I fear?" In the space below, brainstorm ways to manage the safety risks of the consequences you identified for each factor. Place a checkmark next to the ones on the following list and use the extra lines to write in other ideas. As before, do your best to identify reasonable, realistic steps you can take to minimize risk.

- ☐ I will refuse to go into a private space one-on-one with anyone.

- ☐ I will ask someone to walk me to my car when I'm ready to leave.

- ☐ I will leave the party early, by 9 p.m.

- ☐ I will watch any drink I have at all times.

- ☐ I will tell a family member or friend about my whereabouts, and perhaps turn my location on, on my phone, so that they know where I am.

☐ _____

☐ _____

☐ _____

Revisit each factor and, given the mitigation strategies you've identified, ask yourself how unsafe those factors now feel, on the same scale (1 to 100). Have they changed at all? What do you notice? While your "unsafe score" may not lower for all factors, you may find that you're able to sufficiently address some of the factors to a point where the overall situation feels more doable to you. For example, although you may have to attend the party alone, and while maybe it doesn't make you feel any safer to have a friend to check in with while you're there, perhaps you feel a sense of relief when you tell yourself that you'll leave the event by a decent time or that you'll keep an eye on your drink. Although not all factors, or subfactors, can be solved, when you can address some of them, you may feel a bit less unsafe overall.

_____

_____

_____

Finally, before making a final decision to enter the situation or not, you may want to get the perspective of other people whose judgment you trust. Is going to a party alone something others in your peer group tend to do? When and when not? Under what circumstances? Is this a relatively safe situation for most people or is it generally ill-advised, such as hanging out in a dark parking lot alone in a crime-ridden area of town? Once you've done your own analysis, feel free to bounce it off of others!

_____

_____

_____

TIPS FOR CONTROLLING WHAT YOU CAN

Practice this technique whenever you are considering a situation that is potentially unsafe or that just feels unsafe. Through the process, you can confirm what your body and amygdala are telling you (Danger! Danger!), or you may conclude that the alarm bells reflect an old trauma response and not the true dangerousness of the situation. Seeking outside perspectives can further help you clarify the difference. If or when you've determined that a situation is actually relatively safe, and meaningful to you, the next step is to get out and do it! For your brain to learn that something is not dangerous, and for you to feel safe in that situation, you need to actually do it. It's through reengaging with safe people and situations that the brain comes to realize that those people and situations are in fact relatively safe. With practice, this becomes easier.

A Pause to Reflect

You've reached the end of this chapter, which means that it can be helpful to take a moment and reflect on the skills, both from this chapter and from others, that may be helpful to incorporate when you feel as though safety- and control-related issues are causing problems in your life. Some of these skills are bottom-up, focusing more on working with the body and breath in order to reduce amygdala activation. Other skills emphasize cognitive work, using the prefrontal cortex, which can help you slow down and reason through which types of strategies might work best to keep you safe. The following worksheet can help you organize your thoughts by listing the exercises that you've either found helpful or would like to try in the future. You can also visit http://www.newharbinger.com/48848 to download a copy of this exercise.

Skills for Managing Safety- and Control-Related Concerns

The following chart is a reminder of the skills in this chapter and a place to record how each skill works for you. When you notice safety- and control-related concerns are negatively impacting your life, choose a skill that is appropriate for the situation. The second column tells you when the best time to practice is. In the third column, make a note of when and how you try the skill over the coming week. Then, take a moment to reflect on your experience. If you have learned other useful skills or techniques that help, add them to the chart in the blank spaces. If you need more space, you can print more pages or keep a journal to reflect on your experiences.

Skill/Technique	Best Time to Practice	When I Practiced the Skill (Day and Time as Well as Description of the Situation)	Reflect on the Experience. Was the Skill Useful?
Accessing Control in the Body and Mind	When I need to feel more in control.		
Resource Tapping with Affirmations	When I feel unsafe or worried.		
How to Spot Safe People	When I want to determine if I can feel safe with someone.		
Controlling What You Can	When I feel anxious about not having control in a situation.		

CHAPTER 7

Difficulties with Trust and Relationships

Ever since Brian, who served in the army, returned home from Iraq in 2006, he has had a hard time getting close to people or trusting them. He says, "Out there, I couldn't trust anyone. Even though my friends told me they had my back, I knew that something bad could happen at any time, and no one would be able to save me. I saw this all the time, people getting hurt or killed, and no one could help. I learned then that the only person you can depend on is yourself."

Since his discharge, Brian's family and friends have complained that he seems different, and not like himself. One of their primary concerns is that while he used to be the "life of the party," extroverted, and fun-loving, he now isolates, tends to have heated conflicts with others, and won't open up to anyone. When confronted about this, Brian admits that he has lost most of his friends and he no longer feels like he can trust anyone or get along well with others. While he still communicates with his family, it is only because they insist on calling him once per month. However, he refuses to attend most family gatherings. Additionally, his romantic life has suffered; his wife filed for divorce in 2008, and since then he has only had short-term romantic partners because "getting close to people causes anxiety." Thus, Brian reports that he often "causes a fight" that will end his dating relationships so that he does not have to tolerate the risk and discomfort of closeness. This leaves him feeling lonely and frustrated.

Check-In: Your Experiences

After reading about Brian, take a moment to think about your experiences. You may choose to reflect and journal on the following questions:

- Have you ever felt similar to Brian? Perhaps feeling as though you can no longer get close to people or trust others? What's that been like for you? For your relationships?

- Have you been told that you seem different after a terrible experience? What did others notice about you? What did you think about what you were told?

- Have you noticed, after experiencing a betrayal or traumatic event, that you've cut off from relationships or taken steps to distance yourself from people you used to be close to? How does it feel when you do that? Does it feel safer? Do you feel lonely? Or do you feel apathetic about it at times?

How Relationship and Trust Difficulties May Impact Your Life

Relationships are a central part of life and are important for both physical and mental well-being. Experiencing chronic difficulties in this domain, including mistrust, is associated with high cortisol (the stress hormone), anxiety, depression, loneliness, and loss of meaning in life. In Brian's case, his disintegrated relationships reportedly contributed to his PTSD and depression, making him feel worthless and alone.

Have your relationships changed since your trauma? Do you find it hard to connect with and trust people? Relationship challenges are extremely common after trauma—the reactivity and stress you feel, as well as betrayals you may have experienced in the past, make it hard to get close to people the way you might like. However, there are things you can do to improve your connections with others. Take the following self-assessment to better understand how post-trauma symptoms may be impacting your relationships. You can also visit http://www.newharbinger.com/48848 to download a copy of this exercise.

Relationship Difficulties Self-Assessment

Read the following and circle the best response from 0 to 3, where 0 = None/not at all, 1 = A little bit/sometimes, 2 = A moderate amount/often, and 3 = A lot/most of the time. Since the trauma(s), do the following statements resonate with you?

It is difficult for me to trust strangers or new people.	0	1	2	3
It is difficult for me to trust loved ones and longtime friends.	0	1	2	3
I am not easily able to make new friends or I tend to lose relationships.	0	1	2	3
My romantic relationships are unstable.	0	1	2	3
I find excuses to avoid seeing or talking to people that I know I like or love.	0	1	2	3
It feels anxiety-provoking to get close to people.	0	1	2	3
I prefer relationships that are emotionally distant—relationships in which I don't have to share or talk about my feelings.	0	1	2	3
My relationships tend to have a lot of conflict.	0	1	2	3
I have a hard time getting along with others.	0	1	2	3
I tend to choose the "wrong people" to attach to.	0	1	2	3
People in my life don't understand my experiences or difficulties.	0	1	2	3
I know I love my family and friends, but I have a hard time accessing loving feelings toward them.	0	1	2	3
I am rarely the one to reach out in my relationships.	0	1	2	3
Sometimes I agree to doing things with a loved one or friend, but then back out at the last minute.	0	1	2	3
I ghost people.	0	1	2	3
It makes me very uncomfortable to depend on others.	0	1	2	3
People can't be trusted.	0	1	2	3
People only want to be connected to me because of what I can give them or do for them.	0	1	2	3
I prefer to be in control in my relationships.	0	1	2	3
If I don't hear from someone, I assume something bad about them (for instance, that they hate me).	0	1	2	3
People will hurt me if given the chance.	0	1	2	3
I feel suspicious of others.	0	1	2	3
Even my closest family and friends would lie to me.	0	1	2	3
Total				

Add up the total score from the above items, and record them next to "Total." A score of 24 or higher indicates significant relationship difficulties and suggests you may benefit from working on your relationships and ability to trust others.

This assessment details the most common relational challenges individuals experience after trauma, especially if the traumatic event involved deceit, betrayal, malintent, or harm. You may have a hard time getting and staying close to people and find it difficult or impossible to trust others. This can even include longtime romantic partners, family, and close friends. Additionally, you may find that your relationships contain more conflict after the trauma, and sometimes the arguments can feel extremely intense and out of control. In Brian's case, heated conflicts, which often led to breakups and estrangement, were a primary way he pushed others away.

Use the following exercise to reflect on your self-assessment score. You can also visit http://www.newharbinger.com/48848 to download a copy of this exercise.

Reflect on Your Self-Assessment Score

What did you score on the self-assessment? Was this surprising to you? Take a moment to reflect on, and process, what you just learned about yourself. You may consider the following questions as you reflect on your experience with this.

Was your score higher than you thought it would be? Lower?

_____

_____

What does your score indicate about your relationships and how you relate to others?

_____

_____

_____

Were there items on the assessment you hadn't considered before?

_____

_____

_____

Were there items that surprised you?

_____

_____

_____

How do you think these symptoms have been impacting your life more broadly?

_____

_____

_____

Why Relationships Are Difficult After Trauma

Although trauma can impact people in very different ways, it is common to experience relationship difficulties if the trauma(s) you suffered involved harm caused by other people. For instance, rape, assault, abuse, neglect, betrayal, and other types of trauma may lead you to mistrust people. It may also be difficult to get close and attach to other people in healthy ways. Sometimes you may end up choosing partners that are untrustworthy as a way to avoid closeness ("If I know he's dishonest, I won't have to get very close to him and can continue to be emotionally distant…"). Other times, you may choose healthy partners or friends but feel so anxious about closeness that you sabotage the relationship. Thus, after trauma, you may remain in toxic relationships and avoid healthy ones.

While this initially seems illogical, it actually makes sense: it's hard to create loving relationships when people have hurt you in the past! Much like the brain can learn to be on guard and anxious after a trauma to keep you safe, it can also learn to mistrust people and feel anxious about attachment. This, too, is ultimately meant to keep you safe. In essence, the brain has learned that other people can be dangerous and can hurt you, so to be safe, it's best to avoid becoming vulnerable with, or dependent on, other people. Remember, the brain's main objective is to keep you safe and alive, not happy and connected! Thus, in order to have healthy relationships, you have to be able to learn (or relearn) how to feel safe in a relationship context. This was a large challenge for Brian, as he repeatedly experienced how untrustworthy and hurtful people could be.

To better understand how relationship difficulties develop after trauma, especially trauma that involves other people's actions (as opposed to a natural disaster or other non-man-made event), here is a summary of what many people experience.

1. **Trauma happens, and the amygdala activates.** When you experience trauma, your amygdala, the threat detection center of your brain, activates. This activation signals that something terrible is happening and makes you feel afraid, out of control, and overwhelmed.

2. **Your hippocampus encodes the memory of the trauma.** After the amygdala activates, it sends signals to other parts of the brain, alerting them that something bad and dangerous is happening. Because the traumatic event is very distressing, the main long-term memory center of the brain, the hippocampus, immediately attempts to activate as it works hard to encode the memory. Because the brain's main goal is to promote your survival, the brain as a whole, and the hippocampus in particular, wants to remember the events, people, etc. that have hurt you in the past. If you can remember things that have almost killed you, you can avoid them in the future. Thus, during the trauma, the hippocampus struggles to create a memory of the most salient parts, so that when you encounter those elements in the future, your brain will send you danger signals, saying, "Run, run! This has been dangerous in the past!"

3. **The hippocampus reminds you of your trauma in the form of a "trigger."** When the traumatic event is over, the hippocampus continues to encode and consolidate the memory of it. Though the details of trauma memories are often fuzzy, fragmented, out of order, and unclear, the general "gist" of the memory becomes burned into the hippocampus for many people. What this means is that while it might be hard for you to recall the context or details of a traumatic event, your brain has held tightly to the *experience* of the event, how it felt, and the general elements it included. If the event was a sexual assault, for instance, the brain may not encode it as a coherent, clear movie. Rather, it may retain certain components of the memory and draw firm, black-and-white conclusions about what the event means. Here's an example of what this encoding might look like:

> *"I was hurt. Men can't be trusted. There was a brown carpet. The lamp was broken. The sound of a barking dog. Sick to my stomach. A sharp pain in my leg. He tricked me. My phone didn't work. No one was home. We studied physics together. We were alone. Never trust men."*

As you can see, the brain will not necessarily encode a memory clearly and in order, and context is often missing. The "never trust men" conclusion may not be rational, but it's what this person's brain has concluded to be *the safest policy to adopt* after a trauma. And what's ultimately safest (in the most extreme sense) will not necessarily be what's healthiest for you if you are in a low-risk, relatively benign environment. When the brain encodes traumatic memories in a generalized manner, like this, it means that it will signal the alarm bells to go off any time you get close to, or perhaps every time you even encounter, a man. These alarm bells, which scream, "Get away from men; they are not safe!" make it so that men become a trigger for you. And as a result, from that point onward, it may become especially difficult for you to have positive relationships with men.

4. **Your amygdala and hippocampus team up to keep you "safe."** When other people become a trigger, thanks to the hippocampal encoding of the trauma, the amygdala jumps in and offers to help strengthen the emotional "punch" of the

trigger, usually to motivate you to stay away from it! Specifically, the amygdala learns that when the hippocampus encounters something that is reminiscent of the trauma, it means it's dangerous, and so the amygdala activates (remember, the amygdala is the threat detection center). So, the hippocampus may think that a new man in your life is dangerous because men have hurt you before. When this happens, the hippocampus activates and warns you that this man may be dangerous, based on past experience. The amygdala, which is a neighbor of the hippocampus, receives this information and responds by setting off the danger alarms in your brain and body. When this happens, you may feel tense, antsy, and anxious. What's sneaky about this, though, is that the hippocampus and amygdala often do not tell you exactly why you're feeling this way. In other words, it's actually somewhat rare that your brain will say to you, "You're feeling anxious about this man because you were assaulted by another man." Instead, you may only be told by your brain, "This man is dangerous and no good. You need to run. This feels wrong." This means that people often react to subtle trauma reminders in a way that is not conscious. Over time, this can make it hard for trauma survivors to trust their own instincts.

5. **Relationships suffer.** When the brain reacts in the way it does after relational trauma, it makes it difficult to have healthy, close connections with others. How can you enjoy relationships when they always seem to set off alarm bells? Thus, you may tend to choose unhealthy partners or avoid healthy relationships. You may also mistrust friends and family and prefer to isolate in order to get the alarm bells to silence. While this may feel good in the short term, over time it can contribute to (and worsen) mental health issues such as PTSD, depression, and anxiety.

Now that you better understand what can happen in the brain after trauma and why relationships can be impacted, use the following exercise to reflect on how your own brain might be working in ways that keep you disconnected from, or mistrusting of, others. You can also visit http://www.newharbinger.com/48848 to download a copy of this exercise.

Your Relationship Difficulties

Answer the following questions to reflect on your relationship difficulties.

What my brain tells me about relationships in general:

_____

_____

_____

What my brain tells me will happen if I trust people:

_____

_____

_____

What my brain tells me if I start to get close to someone:

_____

_____

_____

Ways I might push people away because of what my brain tells me:

_____

_____

_____

Examples of times my relationship "alarm bells" tend to go off:

_____

_____

_____

How I want to respond to my brain when the "alarm bells" sound (check the responses that resonate most with you):

☐   Thank you for keeping me safe, brain, but I think we are okay in this situation.

☐   Brain, I appreciate your efforts, but right now they are not helping me.

☐   I understand why you're acting like this, brain, but right now I don't want to respond like this.

☐   Brain, I hear the alarm bells you're ringing, but we are not going to act on them right now.

☐   _____

☐   _____

As you think about how to respond to what your brain is telling you, consider how to do this in a compassionate, gentle way. When you are harsh with yourself or self-critical, it activates the fear center of the brain, the amygdala, which just makes you feel worse. Ways I can be gentle and compassionate with myself:

_____

_____

Terms Your Therapist Might Use to Explain This Challenge

If your experiences are similar to Brian's, a therapist may use the following descriptors and terms in your discussions, or in your therapy notes:

- **Cognitive rigidity.** This refers to how you think about events and how you tend to view yourself, others, and the world. If you experience cognitive rigidity, it means that you often interpret things as black and white and have a hard time shifting your interpretations or perceptions. For instance, you may conclude that "no one can be trusted" and feel resistant to considering that it's possible to trust others to varying degrees. Instead, you think of trust as "all or nothing": people can be trusted, or they can't be.

- **Negative alterations in cognition and mood symptoms.** This is a *DSM-5* PTSD symptom cluster that describes changes in mood and thinking that often occur in PTSD. Some symptoms include feeling guilt or blame after a trauma, adopting negative or "all or nothing" viewpoints or assumptions about yourself or others, difficulty experiencing positive emotions, and withdrawing from others and activities you used to enjoy.

- **Emotional alienation.** This term describes a common phenomenon after trauma wherein a survivor pulls away from others, including important relationships. This "alienation" can present as you not returning phone calls or texts, canceling plans, or not showing up to important events. Or you may be physically present but emotionally absent. Those around you may complain that you no longer seem interested in the relationship, and you may seem uncaring or emotionally disconnected. Even relationships that have been close and intimate for many years may suddenly feel distant and detached.

What Your Brain Needs

Human relationships are believed to be central to survival, which is a good thing, because it means that your brain is wired for connection! That's a great starting point to remember: your brain ultimately wants you to be attached to other humans. To relearn safety and comfort in relationships and build trust again, here is what your brain needs.

1. Bottom-up techniques that reduce activation of the "smoke detector" of the brain (amygdala). These techniques can help you learn how to feel less stressed and reactive when your brain sends you incorrect signals about the dangerousness of others.

2. Top-down techniques that activate the "memory storage unit" of the brain (hippocampus). These cognitive skills can help you reason through and shift some of the thoughts and beliefs about others that you may have developed after trauma.

3. Behavioral techniques that help the hippocampus and amygdala relearn that (some) relationships are not dangerous. These skills encourage you to slowly reengage with relationships in structured and contained ways that help your brain readjust to having close and loving relationships.

The next section details four bottom-up, top-down, and behavioral skills that can help you learn how to safely engage in and improve the quality of your relationships. Additionally, skills from chapter 1, which focus on ways to reduce amygdala activation to help you feel less anxious, may also be beneficial as you work to improve your relationships. When you are able to reduce your amygdala's activation and teach the hippocampus that not all relationships are dangerous, you can relearn how to enjoy attachments to other people.

Skills for Meeting the Challenge

In this section, you'll learn four skills for meeting this challenge: creating a profile of the physical symptoms you experience when you're feeling trustful and mistrustful, to make these experiences easier to recognize and adjust for when they arise; learning to broaden your instinctive perspectives on people and events by interpreting situations realistically rather than reactively; learning to see your relationships, and trust within those relationships, in shades of gray, rather than the black-or-white, all-or-nothing thinking trauma can predispose you to; and implementing changes in your behavior to make the relationships you value less anxious, more trusting, and more beneficial to you.

Skill 1: Creating a Physical Profile for Trust and Mistrust (Bottom-Up)

The experiences of trust and mistrust are largely physical and emotional, and they aren't always based on logic. While cognitive (thought-based) techniques for improving the ability to have healthy relationships and trust can be helpful, insofar as they help promote behavior change, thoughts alone don't usually convince people to trust others! Rather, trust and mistrust are experiences we have, oftentimes based on past experiences, that are felt in the body. In this exercise, you'll create a physical profile for both trust and mistrust, to better understand the distressing sensations you may be feeling when trying to get close to someone (adapted from Sweeton 2019). If you can recognize and work with the distressing sensations as they arise, you can learn how to manage them and reduce the stress associated with closeness.

CREATING A PHYSICAL TRUST PROFILE

Use the worksheet following the directions to map your trust profile. You can also visit http://www.newharbinger.com/48848 to download a copy of this exercise.

1. **Identify a memory of trust.** Bring to mind a memory of someone you have trusted in the past. Perhaps this was in childhood, with a caregiver, teacher, or friend. If you've never trusted anyone, you may replace the word "trust" with "comfort," thinking instead about someone you've felt comfortable with. Close your eyes and focus on the person involved, trying to picture them as clearly as possible in your mind's eye. What did they look like? How did their voice sound? What were they saying or doing at the time to make you feel trusting or at ease? Take a moment and stay with the memory of this person.

2. **Connect with the emotion of trust.** Keeping this person in mind, begin noticing what this experience of trust feels like in your body. Feel into your whole body and make a mental note of anything you initially notice, such as warmth or relaxation. Stay here for a minute, just noticing any physical experiences or sensations.

3. **Focus specifically on surface sensations.** As you keep this trustworthy person in mind, ask yourself the following:

a. "As I think about this person, can I feel any pressure on the surface of my body?" Starting at your feet, scan upward, noticing any pressure that may be present. For instance, you may notice that there is pressure on your legs and backside if you are sitting down.

b. "As I think about this person, can I feel any sensations on the surface of my body?" External sensations may include tingling, itching, or other experiences.

c. "As I think about this person, can I feel the temperature of my body?" Starting at your feet, scan upward, noticing the temperature of various regions of the body and noting how they differ (for instance, your feet may be cooler than your stomach).

4. **Focus on internal sensations and experiences.** As you keep the trustworthy person in mind, ask yourself the following:

a. "As I think about this person, can I feel any sensations on the inside of my body?" Internal sensations may include stomach rumbling, head pulsating, "butterflies," etc.

b. "As I think about this person, can I feel the temperature inside my body?" To notice internal temperature, consider focusing on your abdomen and extremities, feeling into these areas and noticing feelings of warmth that might be present.

c. "As I think about this person, can I detect any relaxation or tension in my muscles?" Some areas of the body tend to hold tension, such as the neck, back, and jaw. However, when feeling relaxed and happy, some areas may feel particularly relaxed, such as your arms, legs, face, and abdominal region.

5. **Map out the sensations of trust.** Open your eyes, still accessing the memory of this person and noticing the sensations associated with trust (or comfort). While you are still experiencing some of these sensations, jot them down on the following worksheet, noting where in the body you experience them. You may use this profile as a resource, to help you feel calmer during times of stress by accessing this person/memory, and to remind yourself that trust may not always feel scary.

Physical Trust Profile

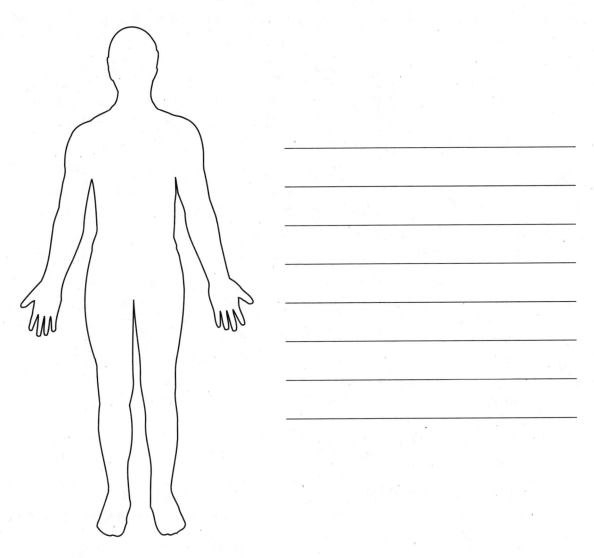

CREATING A PHYSICAL MISTRUST PROFILE

Use the worksheet following the directions to map your mistrust profile. You can also visit http://www.newharbinger.com/48848 to download a copy of this exercise.

1. **Identify a memory of mistrust.** Bring to mind a memory of a time (but not a trauma!) when you felt betrayed or like you could no longer trust a person. Close your eyes and bring an image of this situation and person to mind. Try to access how you remember the person looked, sounded, and what they did. Take a moment and stay with the memory of this person. If this person or memory is linked to a traumatic event, take a bit of a different approach. Instead of imagining the actual event and person, bring to mind a point in time after the event where you thought about or talked about this person. Thus, instead of imagining the trauma directly, you're bringing to mind a memory of yourself thinking or talking about this individual afterward. If you do not possess such a memory, simply bring this untrustworthy person to mind, imagining them alone, outside of any particular memory.

2. **Connect with the emotion of mistrust.** Keeping this untrustworthy person in mind, feel into your whole body and make a mental note of anything you initially notice, such as tension, trembling, or sickness. Stay here for a minute, just noticing any physical experiences or sensations.

3. **Focus specifically on surface sensations.** As you keep this untrustworthy person in mind, ask yourself the following:

   a. "As I think about this person, can I feel any pressure on the surface of my body?" Starting at your feet, scan upward, noticing any pressure that may be present.

   b. "As I think about this person, can I feel any sensations on the surface of my body?" External sensations may include tingling, itching, or other experiences.

   c. "As I think about this person, can I feel the temperature of my body?" Starting at your feet, scan upward, noticing the temperature of various regions of the body and noting how they differ (for instance, the head may feel warm or hot).

4. **Focus on internal sensations and experiences.** As you keep the untrustworthy person in mind, ask yourself the following:

   a. "As I think about this person, can I feel any sensations on the inside of my body?" Internal sensations may include feeling dizzy, rapid heart rate, or shallow breathing.

   b. "As I think about this person, can I feel the temperature inside my body?" To notice internal temperature, consider focusing on your head, abdomen, and extremities, feeling into these areas and noticing any temperature changes.

   c. "As I think about this person, can I detect any tension in my muscles?" Some areas of the body tend to hold tension, such as the neck, back, and jaw.

5. **Map out the sensations of mistrust.** Open your eyes, still accessing the memory of this person, and notice the sensations associated with mistrust. While you are still experiencing some of these sensations, jot them down using the following worksheet, noting where in the body you experience them. As you do this, make a conscious, mental link between these sensations and the experience of mistrust.

Physical Mistrust Profile

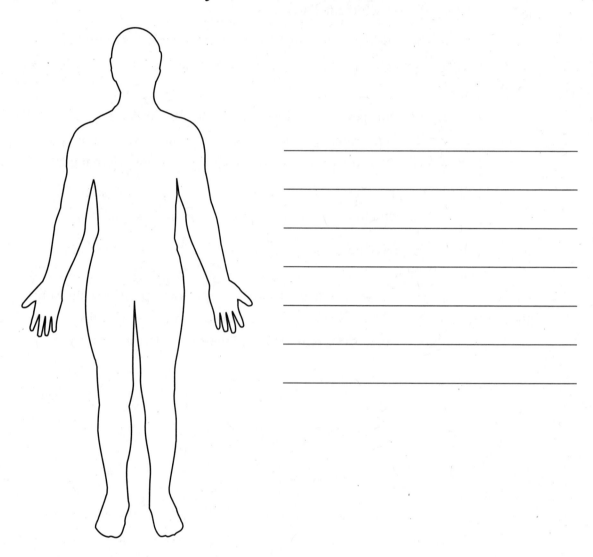

TIPS FOR CREATING A PHYSICAL PROFILE FOR MISTRUST

The next time you're in a social situation or around friends, partners, or family, intentionally check in with yourself to see whether you're getting "triggered," meaning you notice one or more of those sensations. If your body responds to social contexts by producing some components of your mistrust profile, you may feel uneasy in these situations. This may lead to you pulling away, pushing others away, initiating arguments, or engaging in other stress-based sabotaging behaviors.

However, if you can begin to notice these physical changes in real time, you may be able to move your attention inward, focusing on and managing the sensations directly, instead of reacting unconsciously to them in ways that prevent you from staying connected to others. For example, if on a date you notice that your head feels hot, your breathing is shallow, and you feel tense in your shoulders, you may be able to say to yourself, "This is what mistrust feels like. Because I don't really know yet whether this person is untrustworthy, these reactions might be based on my past trauma. It is okay and understandable that my body is responding in this way. I'm not going to act on these sensations right now, though." Then, instead of running away from the date, you may focus for a moment on taking deep breaths or briefly excuse yourself to ground and recenter using anxiety-reduction tools (see chapter 1).

Noticing and managing mistrust-related sensations, without acting them out or assuming they mean that the person you're with is not to be trusted, is a healthy way of "managing your baggage." As the saying goes, having baggage is fine, as long as it's labeled! While it's true that sometimes these sensations can indicate red flags or a gut feeling that means a person is dangerous, that may not always be the case. In particular, if you tend to feel these sensations even with friends and family who have never betrayed you, or with new people you're seeing in safe contexts who have not harmed you, these danger signals may not be accurate or helpful and may instead be trauma reenactments where you become triggered in the absence of real danger. This in turn can prevent you from experiencing close and fulfilling relationships with safe, trustworthy (or relatively trustworthy) people.

Skill 2: Broadening Your Perspective (Top-Down)

Broadening your perspective (Sweeton 2019) is a top-down, cognitive technique that teaches you to think differently about other people and events by incorporating additional information, besides the information that comes most readily to mind, to help reduce the distress you feel. This information can include facts about the person or event (who was present, when it took place, etc.) as well as speculations or opinions. Sometimes after trauma people adopt black-and-white thinking, where they believe the world is all bad or no one can be trusted. However, usually the truth is somewhere in between, and this exercise helps you take a more balanced position on topics such as trust.

Use the following worksheet to help broaden your perspective on a person. You can also visit http://www.newharbinger.com/48848 to download a copy of this exercise so you can complete it with multiple people.

Broadening Your Perspective

Write down a few sentences about a distressing event that occurred recently with another person. Be specific about the facts of the event, such as what happened, who was there, where it took place, and when it took place. Refrain from interpretations or speculations about the causes of the event right now and focus just on the facts.

The Facts: _____

_____

_____

_____

Next, write down a few sentences about *why* you think the event occurred and how you interpret the cause and meaning of the event. This simply reflects *your* perspective and may or may not be entirely accurate.

Why It Happened: _____

_____

_____

_____

Identify the "main character(s)" of this event, besides yourself. Who was the main person, or main people, who you believe caused the event to happen? Just list them below.

Main Characters: _____

_____

_____

_____

Now, go back to the first two steps and reread what you wrote about the facts of the event and your interpretations of the event. When you think about the event like this, and the main character(s) involved, what emotions do you feel? What emotions did you feel at the time the event was happening?

My Emotions: _____

_____

_____

_____

With the event still in your mind, imagine that you are the main character (or one of the main characters) of that event. Pretending for a moment that you are that person, taking a third-person perspective, begin to speculate why the main character may have acted the way they did. List at least five explanations for the main character's behavior *that do not reflect poorly on the main character or imply malicious intent toward others* (including you). These are not facts about the situation; rather, they are speculations about the cause of another person's behavior that are either neutral or positive. Next to each explanation, jot down how much you "buy into" it as being a likely reason, using a percentage (0 to 100 percent). It is okay if you do not believe these explanations are true, but assign a "validity percentage" to each one.

Alternative Explanations: _____

_____

_____

_____

_____

Imagine that the main character was in fact experiencing one or more of the things you just listed in step 5 (the neutral or positive explanations). What emotions do you feel now, with this broadened perspective and additional information? Is the person any more trustworthy when you imagine their reasoning to be less negative?

My Emotions: _____

_____

_____

_____

Make a list of all the reasonable causes of the main character's behavior that you can believe to a 30 percent degree or higher, including your initial negative interpretation and any neutral or positive explanations. Refer to step 5 to remember which explanations you rated as 30 percent or higher. When you read all these explanations together, notice whether it shapes your overall perception, or at least broadens it. Even if the likelihood of the negative interpretation is 80 percent true, it may help you feel less distressed if you also realize that you hold a 35 percent belief of a neutral explanation *also* being true. Oftentimes, multiple factors play a role in people's behavior, as opposed to just one. And it is possible that not all motivators are purely negative.

List of All Reasonable Explanations: _____

_____

_____

_____

TIPS FOR BROADENING YOUR PERSPECTIVE

It's important to consider the possibility that not all motivations are negative, as it is our interpretations about *why* things happen that determine our emotional reactions to them and how much we trust or distrust people. Events themselves do not "make" us feel bad or good; rather, it is how we think about them that creates suffering or contentment. This cognitive reappraisal exercise broadens your perspective and allows you to "step into others' shoes" to imagine how others' behaviors may make sense. This is not done in order to excuse the behavior, but to feel less distressed about it and to expand the capacity to trust others (even if just a little). It is much easier to trust someone with positive, or even neutral, intentions than it is a person with 100 percent malintent! Keep in mind that while the cognitive reappraisal is based on speculation only, *all* interpretations are speculative anyway, and you rarely know the entire truth about any event or circumstance. People are rarely completely trustworthy or untrustworthy—it's often somewhere in between and situation dependent.

Skill 3: Seeing Trust in Shades of Gray (Top-Down)

After trauma, it's common for people to think of trust in a black-and-white way, where others are either fully trustworthy or not at all trustworthy. When trust is thought of in this way, all it takes is one small question or error and people become exiled to the "untrustworthy" category. This can be problematic, as it makes it difficult to keep connections to others because most people make mistakes at some point during a relationship. Being rigid in how you define trust can be a sort of wall you build to keep yourself isolated and unable to attach to other people. If this is something you struggle with, you may consider rethinking trust and the varying ways someone can be trusted or distrusted. Even if you can't trust a person in one way, or for one thing, you may be able to trust them for something else. The following list outlines some different areas, or domains, in which a person might be trusted (or mistrusted), though the list is not exhaustive. Notice how being trustworthy in one of these areas doesn't necessarily mean the person would or would not be trustworthy in the other areas!

- Trust with your life (or trusting that the person will not kill you).

- Trust with your physical safety (or trusting that the person will not physically assault you).

- Trust with your emotional safety (or trusting that the person will not use something against you in an attempt to emotionally harm you).

- Trust with your finances (or trusting that you can loan money and know it will be repaid, or being willing to assign power of attorney to someone to manage your finances).

- Trust in another's fidelity (for instance, trusting that your romantic partner will not be unfaithful).

- Trust in another's loyalty (for instance, trusting that friends or partners will support you and not speak badly about you to others).

- Trust in another's reliability (for example, trusting that they will show up to events they've committed to or will return your phone calls).

- Trust in another's emotional stability (such as trusting that the person will be able to control their anger, maintain emotional consistency, and remain respectful even when upset).

- Trust in another's honesty (for example, trusting that the person will be honest about how they feel, what they're doing, etc.).

- Trust in another's word (for instance, that they will keep their promises).

TIPS FOR SEEING TRUST IN SHADES OF GRAY

As you look at this list, you may think of different people as you read each item. You may remember people from your life you felt you could *not* trust in certain areas, or maybe people who have been particularly trustworthy in some area(s). That is a common experience, and more normal than not; people are rarely always trustworthy or always untrustworthy. What's important is that you build and maintain relationships that fulfill the trust domains you value the most. And of course, with some types of relationships, such as a marriage, the standards for trust will be higher than, say, with a loose acquaintance—so keep realistic expectations in mind, too, as you think about how your experiences with trauma affect your ability to trust others in the present day.

Additionally, as you look at this list, it can be helpful for you to consider whether *you* are trustworthy in different domains and whether your own trustworthiness varies depending on your relationships. If you behave in ways that are trustworthy, it's more likely that you will find others who are trustworthy as well, though sometimes it takes time to really know how trustworthy a person is in different domains.

Skill 4: Exposure Therapy for Trust (Behavioral)

The previous skills in this chapter have focused on understanding your physical reactions to trust and mistrust, redefining trust as being more nuanced and multifaceted, and considering how interpretations of others' behaviors may contribute to feelings of mistrust. These skills, in addition to anxiety-reduction techniques, may be helpful for you to practice alone.

However, while attachment ruptures, trauma, and betrayal all occur in the context of other people, healing also happens in the context of other people. Some people believe they need to isolate from others until they can "fix themselves" and "heal," but this is often less effective than anticipated. Thus, skills that are practiced alone to improve relationships can be helpful to an extent, perhaps as a way to learn how to recognize and manage anxious feelings or to examine and shift problematic or black-and-white thinking. Still, these techniques should be thought of as preparation for the *real* work, which is getting back out into the world, into your relationships, and practicing the tools in real time, even while triggered! In other words, you have to dare to have relationships in order to learn how to have relationships.

You can think of getting back out into your relationships as a type of "exposure therapy," which means that you're intentionally doing something uncomfortable and anxiety-provoking, and facing a fear, in order to overcome it. With time, as you continue to reengage in relationships, you can begin to feel more at ease in them. Without practice, your feelings of discomfort won't improve. However, a key strategy for doing exposure therapy well is to titrate into your feared situation, meaning you *slowly* allow yourself to get close to others.

For example, you might begin by trusting those around you with very small, inconsequential pieces of information or requests. So, instead of telling your entire life story to someone on a first date, see whether you can trust them with small things, such as calling you after the date if they say they will. While it might be tempting to quickly conclude that someone is untrustworthy, try to tolerate the distressing feelings that arise when you first meet someone and see how they behave over a few hours, days, or even weeks before drawing a firm conclusion (unless they do something blatantly untrustworthy, such as not showing up to the date!). Moving slowly into trust will minimize the overwhelm you might feel if you were to offer too much trust too soon, but will likely still feel uncomfortable enough for you to grow.

Following are some ways you might offer others small amounts of trust, to see whether you will be able to form a closer relationship to them. These small trust offerings can be used with new people in your life or with people who have violated your trust in the past (if you want to begin rebuilding trust with them). They can also be practiced with people who have never betrayed your trust but whom you have a hard time trusting simply because of the trauma you suffered.

- Call the person, and see if they return the call.

- Offer to meet the person and see if they show up, and on time (or, if they are more than ten minutes late, make a note of whether they reached out to tell you they'd be late).

- Disclose one piece of personal (but harmless) information about yourself and see how the person reacts. Then, wait some time and see whether it seems like they've told other people what you disclosed or whether they're willing to use it against you when angry. You may not know this right away, but you may be able to determine how loyal they are over a long period of time.

- Pay attention to how the person talks about other people. Are they betraying others in how they are talking to you about them? If they are willing to talk about others negatively, and the other people don't know about it, they may be willing to do this to you, too.

- Make a note of how they handle conflict and whether they remain respectful, consistent, and trustworthy even when angry.

- Ask the person for a small favor, and see if they are willing to follow through on it.

- Ask the person about their own vulnerabilities, weaknesses, and mistakes, to see if they are able to be accountable and honest.

- Notice over time if the person's words and actions align.

TIPS FOR EXPOSURE THERAPY FOR TRUST

As you can see, some of these recommendations require time. This is why it's best to move slowly into trust and to view pressure from another person to trust them quickly as a red flag. After trauma, you may need to build trust very slowly and at your own pace. As long as you're moving, you're progressing; no need to speed along with this! Moving slowly, titrating into trust, will allow you to desensitize in bite-size chunks to the distressing sensations, thoughts, and emotions associated with mistrust. This will eventually help you begin enjoying the experience of trusting and feeling close to others.

A Pause to Reflect

Now that you've come to the end of this chapter, it is a good time to pause and reflect for a bit on what you've learned and the skills that you've found helpful. If you notice that your relationships are stressful, unfulfilling, or distant, remember why this can happen and what you can do about it. The following worksheet can help you organize your thoughts by listing the exercises that you've either found helpful or would like to try in the future. This can include skills taught in this book as well as ones you may have learned from other sources. You can also visit http://www.newharbinger.com/48848 to download a copy of this exercise.

Skills for Managing Relationship Difficulties

The following chart is a reminder of the skills in this chapter and a place to record how each skill works for you. When you notice relationship difficulties are negatively impacting your life, choose a skill that is appropriate for the situation. The second column tells you when the best time to practice is. In the third column, make a note of when and how you try the skill over the coming week. Then, take a moment to reflect on your experience. If you have learned other useful skills or techniques that help, add them to the chart in the blank spaces: If you need more space, you can print more pages or keep a journal to reflect on your experiences.

Skill/Technique	Best Time to Practice	When I Practiced the Skill (Day and Time as Well as Description of the Situation)	Reflect on the Experience. Was the Skill Useful?
Creating a Physical Profile for Trust and Mistrust	When I'm ready to reflect on how trust and mistrust are experienced physically.		
Broadening Your Perspective	When I'm motivated to work on changing my thoughts.		
Seeing Trust in Shades of Gray	When I'm motivated to work on changing my thoughts.		
Exposure Therapy for Trust	When I'm anxious and ready to address feelings of mistrust.		

CHAPTER 8

Isolation and Agoraphobia

After losing her husband in a hit-and-run accident, Susan, age forty-five, was devastated. At first, she found herself staying at home due to profound grief. As time passed, she began to realize that she was afraid to leave home alone, for fear that something bad might happen to her, too. This fear intensified over time, and now she has difficulty ever leaving the house, even when her sister is with her. She feels her life closing in around her and wants to have the freedom she used to enjoy.

Check-In: Your Experiences

After reading about Susan, take a moment to think about your experiences. You may choose to reflect and journal on the following questions:

- Have you ever felt similar to Susan, feeling as though you're becoming fearful of going out as much? Have you noticed yourself isolating more over time? Your isolation might be due to an injury, illness, or a pandemic, where you find yourself isolating long after it is prudent to do so.

- Have other people commented that you seem less responsive or that they've been seeing you less than usual since a very stressful or traumatic event? Or do people tell you that you seem different from how you used to be?

- Do you feel more restricted than you used to, as though it is difficult to go out and fully engage in the activities you used to engage in? Do you find that it is hard to do the things you used to do easily or go to the places you used to go to?

How Isolation and Agoraphobia May Impact Your Life

Agoraphobia, which is a fear of going into places or situations that might cause intense anxiety or panic, can have a profound impact on your life. Even if the anxiety associated with leaving home is not debilitating, and you *can* still do things outside the home, a tendency to isolate can greatly disrupt everyday life. For many trauma survivors, isolation and agoraphobia develop after experiencing a traumatic event. Oftentimes, the survivor avoids situations or places that remind them of the traumatic event. For example, someone who was in a terrible car wreck may avoid driving or riding in a car. Or a trauma survivor of an active shooter may avoid public or crowded places. However, sometimes the person's trauma seems unrelated to the isolation and agoraphobia, such as in Susan's case. Rather, in these situations, the anxiety/agoraphobia may develop simply because, as a result of the trauma and related emotions (fear, grief, etc.), the person has become accustomed to isolating at home. In other words, because staying at home, in their comfort zone, has become a habit, it subsequently becomes difficult to get back out into the public and around people.

The following brief self-assessment, which is adapted from the Panic and Agoraphobia Scale (Bandelow 1995) and UCLA Loneliness Scale (Russell et al. 1978), is a fast way to determine whether you might be experiencing agoraphobia or isolation to an unhealthy extent. You can also visit http://www.newharbinger.com/48848 to download a copy of this exercise.

Isolation and Agoraphobia Self-Assessment

Read the following and circle the best response from 0 to 3, where 0 = None/not at all, 1 = A little bit/sometimes, 2 = A moderate amount/often, and 3 = A lot/most of the time.

I rarely see my friends and family.	0	1	2	3
I find myself socializing much less than I used to.	0	1	2	3
I feel withdrawn.	0	1	2	3
I feel anxious when I'm not at home.	0	1	2	3
I only leave the house when it is absolutely necessary.	0	1	2	3
I feel cut off from the outside world.	0	1	2	3
I am no longer close to people I used to connect with.	0	1	2	3
I am worried I will have a panic attack when I go out or go certain places.	0	1	2	3
I am worried I will feel anxious or uncomfortable when I go out or go certain places.	0	1	2	3
I cannot tolerate being in public places.	0	1	2	3
When I have to go out somewhere, I try to get back home as quickly as possible.	0	1	2	3
I feel that I am totally alone.	0	1	2	3
No one understands what it's like to be in my shoes.	0	1	2	3
If I leave the house, or go certain places, I will experience physical sensations such as shallow breathing, shaking, rapid heart rate, or muscle tension.	0	1	2	3
Sometimes I get so anxious about going out that I cancel my plans.	0	1	2	3
I have missed important events because I am anxious about attending them.	0	1	2	3
Sometimes I don't get the things I need because I refuse to go shopping for them.	0	1	2	3
Public or social events are no longer enjoyable.	0	1	2	3
When I'm not at home, I feel unsafe.	0	1	2	3
Other people have said that I seem isolated.	0	1	2	3
Total				

Add up the total score from the above items, and record them next to "Total." A score of 22 or higher indicates possible significant isolation and/or agoraphobia that may be negatively impacting your life.

Use the following exercise to reflect on your self-assessment score. You can also visit http://www.newharbinger.com/48848 to download a copy of this exercise.

Reflect on Your Self-Assessment Score

What did you score on the self-assessment? Was this surprising to you? Take a moment to reflect on, and process, what you just learned about yourself. You may consider the following questions as you reflect on your experience with this.

Was your score higher or lower than what you expected?

_____

_____

What does your score indicate about how your life may have changed since your trauma(s)?

_____

_____

_____

Were there items on the assessment you hadn't thought about before?

_____

_____

_____

If you're experiencing isolation or anxiety about being out in certain places, is this something you'd like to change? Why or why not?

_____

_____

_____

Why Isolation and Agoraphobia After Trauma Are So Common

When you think about what it means to experience the effects of trauma, you may not think about isolation or agoraphobia, as these aren't discussed as frequently as some other symptoms, such as intrusive memories or hypervigilance. However, it's very common for individuals who come to therapy for agoraphobia to have experienced at least one traumatic event. Sometimes the trauma seems to be directly linked to the isolation or agoraphobia, as in the example of the client who is afraid of driving after a car accident. In these cases, it is clear and understandable why the person wants to avoid cars: their brain has taught them how dangerous cars can be!

For other people, however, there may be no clear link between the trauma and their isolation or agoraphobia. For instance, you may experience something like a tornado or other natural disaster and then find over time that you are increasingly isolated or nervous about leaving the house or being out in public. Although the tornado itself may not seem directly related to your isolation, it's not uncommon for people to develop post-trauma symptoms that don't resemble the actual traumatic event. For some, anxiety about being out, or about leaving the house, can slowly arise simply because the person has become accustomed to being at home. For them, being at home, alone, becomes their comfort zone, and exiting that comfort zone can feel nerve-wracking. Underlying this anxiety may be fears about safety, as in Susan's case, or a worry that people out in the world may be untrustworthy or that something might happen and they could lose control.

To better understand how isolation and agoraphobia occur after trauma, here is a summary of what some people may experience and how these symptoms can develop.

1. **Trauma happens, and the amygdala activates.** When something very stressful or traumatic occurs, the amygdala activates, signaling threat, danger, and possibly catastrophic loss. In turn, the amygdala sends threat signals to other areas of the brain, which then alert the body about the terrible thing that is happening to you (or threatening you). Thus, when trauma happens, it sends shock waves through your brain and body, which is what happened when Susan learned that she lost her husband in a hit-and-run accident. When this happens, the body goes into the stress response, to prepare to fight, flee, or freeze, and the "thinking area" of the brain, the prefrontal cortex, dulls, making it difficult to think rationally.

2. **The "memory center" of the brain struggles to encode the memory.** The hippocampus, known as the "memory center" of the brain, immediately begins to encode the traumatic event as a memory. However, because cortisol, the stress hormone, floods the hippocampus during the experience of trauma, these events are often stored a bit differently. This is because cortisol can compromise the hippocampus's ability to do its job of effectively encoding memories. Specifically, the presence of cortisol in this brain region can make trauma memories feel fragmented and overly generalized (where, for instance, after a car accident with a blue truck you develop a fear of all blue vehicles).

3. **Overgeneralization of fear can lead to isolation and agoraphobia.** When the hippocampus encodes a trauma memory in a way that is overly generalized, it can lead you to fearing things, people, situations, or places that aren't actually usually dangerous. For instance, you may develop a worry, like Susan, that "something bad could happen" or a sense that "the world is a dangerous place." If you begin to believe these generalized statements, it can make it difficult to engage in everyday living because, all of a sudden, nowhere, and no one, feels safe. This can quickly lead to isolation and agoraphobia.

4. **The amygdala activates when you go out or interact with others.** When you begin to believe that it's not safe to go out and do things, or that people can't be trusted, these thoughts can activate the amygdala, signaling more danger. In a sense, you've trained your amygdala to send you fear and danger signals any time you consider going out or interacting with others, which makes doing those things

prohibitive. And the more you avoid those things, the more anxious you'll begin to feel about them over time, and the worse the isolation and agoraphobia will become.

After a traumatic event, your sense of safety, or thoughts about others or the world, may shift. These changes can affect your ability to comfortably engage with others or go out in public. Complete the following exercise to explore how some of your thoughts or beliefs, especially those related to safety, trust, or control, may have changed since the trauma. You can also visit http://www.newharbinger.com/48848 to download a copy of this exercise.

Your Thoughts About Safety, Control, and Trust

Since the trauma, have you noticed that your thoughts or beliefs about safety have changed? For instance, you may think more about safety overall than you used to or it's become difficult to convince yourself that you're safe. Also, some places or contexts that used to feel safe may no longer feel safe to you. What have you noticed, regarding your beliefs about safety, since experiencing trauma?

_____

_____

_____

Some people become worried about feeling out of control after experiencing trauma, and this can be linked to safety concerns. When you don't feel you are powerful enough to keep yourself safe, your anxiety about safety can intensify. This can lead to controlling behaviors, and a fear of being in situations that you can't control, such as crowded places. Since the trauma, have you noticed yourself wanting to be in control and feeling anxious when you're in a situation where you lack control? What has this been like? Has it led you to avoid situations where there is uncertainty?

_____

_____

_____

When traumatic events are caused by other people, it can make it hard to trust others. Many people struggle with trust after experiencing trauma, which is logical because they've often been deeply hurt by other people's malintent or carelessness. Since the trauma, have you noticed that it's become difficult for you to trust others? How is this different from how you used to be? What beliefs do you have about other people's trustworthiness, and how might these thoughts contribute to isolation or agoraphobia?

_____

_____

_____

Answering these questions may clarify some of the reasons you've become more isolated than you used to be or why you feel uneasy about leaving the house. If you've experienced changes in your thinking about safety, control, or trust, your amygdala may be activating whenever you're faced with situations that require experiencing uncertainty, accepting a lack of control, or extending trust.

Terms Your Therapist Might Use to Explain This Challenge

If your experiences are similar to Susan's, a therapist may use the following descriptors and terms in your discussions, or in your therapy notes:

- **Agoraphobia.** This refers to an intense fear of leaving the home, going into crowded places, or going places where you might feel trapped. Agoraphobia is recognized as an anxiety disorder by the *DSM-5*, and it often occurs with panic disorder. It can also develop alongside PTSD and is sometimes conceptualized as an outcome of the repetitive avoidance behaviors and anxiety symptoms of PTSD.

- **Anxiety.** This is a general descriptor that can refer to a particular disorder, such as agoraphobia or panic disorder, or a category of disorders (anxiety disorders). It can also refer to a feeling state ("I'm feeling anxious"), where there is no formal diagnosis present. When anxiety is severe, as it can be in PTSD, it can lead to avoidance behaviors, including isolation from other people.

- **Avoidance symptoms.** Avoidance is one of the key symptom clusters of PTSD, and it includes avoidance of both internal and external reminders of the trauma. Internal reminders include thoughts and feelings related to the trauma (such as terror, racing heart rate, etc.), whereas external trauma reminders include people, places, and situations that are reminiscent of the trauma. If you find yourself isolating and avoiding going to particular places after a traumatic experience, it is possible that your avoidance developed because those situations (or places or people) remind you of the trauma or ways you felt during the trauma. It is also possible that your avoidance is due to thoughts or beliefs you developed after the trauma, such as a belief that nowhere is safe or that no one is trustworthy.

What Your Brain Needs

After trauma, it is common for your brain to conclude that certain (or all!) people, places, or situations are unsafe, uncertain, or untrustworthy. This creates anxiety and often leads to avoidance. However, it's possible to change your brain and learn how to reengage with other people and places. Here's what your brain needs in order for you to reconnect with people, places, and situations you value.

1. Bottom-up techniques that reduce the activation of the threat detection center (amygdala). When the amygdala activates, it sends danger signals to the body and suppresses activation of the thinking center of the brain (prefrontal cortex), making you feel nervous and compromising your ability to think clearly. When amygdala activation is paired with social interactions or going to certain places, you'll likely feel anxious about engaging in those things and may start to avoid them altogether. If your goals are to feel less anxious and avoidant, working on decreasing amygdala activation is important. This is best accomplished through bottom-up, body-focused techniques.

2. Top-down techniques that strengthen the thinking center of the brain (prefrontal cortex) and allow you to better address anxious thoughts related to isolation or agoraphobia. When you practice top-down, cognitive techniques—skills that

require intentional thought and effort—it's like a workout for your prefrontal cortex that helps you think more clearly and logically. As you exercise and strengthen the prefrontal cortex, it becomes easier to address and regulate fear-based thoughts and the resulting avoidance that the amygdala and hippocampus have been perpetuating.

3. Behavioral techniques that help the memory center (hippocampus) relearn a sense of safety, trust, and tolerance for uncertainty. The hippocampus, which helps encode traumatic memories, often stores memories in a way that promotes fear generalization. This can lead to trauma survivors becoming afraid of situations, people, or places that are not objectively dangerous. Exposure-based behavioral techniques, such as slowly engaging in feared activities or going to feared places, are a great way to teach the brain that these situations (or people, etc.) are safe.

The next section details four skills that contain bottom-up, top-down, and behavioral components that can help you address isolation and agoraphobia.

Skills for Meeting the Challenge

In this section, you'll learn four skills that can help you reduce anxiety and reengage with avoided situations, people, or places. These skills include creating a physical profile of fear, imaginal exposure techniques where you'll identify and imagine engaging in feared situations, and a technique that will help you address your avoidance, isolation, and agoraphobia.

Skill 1: Creating a Physical Profile of Fear (Bottom-Up)

In this exercise, you'll create a physical profile of the fear (or general distress) you experience when *imagining that you are engaging* with an avoided person, place, or situation. This will better allow you to understand the physical sensations and experiences that prompt you to avoid that situation. First you will identify the fear and the associated thoughts and feelings. Then you will map the physical sensations caused by that fear. Use the worksheet following the directions to map your own fear profile. You can also visit http://www.newharbinger.com/48848 to download a copy of this exercise to use for multiple fears.

1. **Identify something you've been avoiding.** What is one person, place, or situation that you've been avoiding (either partially or completely) since experiencing a traumatic event?

2. **Connect with what you've been avoiding.** Begin to reconnect with the feared situation in your mind, imagining that you are in that situation, at that place, or with that person. You may close your eyes as you do this. As you connect with your thoughts, begin to imagine your worst fear happening when you're in that situation. For instance, if you've been avoiding elevators because you're afraid of getting trapped in one, imagine that you're in an elevator that has become stuck, and you can't get out. What thoughts come up as you imagine your fear?

3. **Notice your emotions.** As you continue to imagine the avoided situation, notice any emotions you might experience. What emotions are you experiencing while you think of this avoided situation?

4. **Notice your body.** Now, continuing to connect with the emotion tied to these thoughts, shift your attention to your body, noticing any sensation or experience occurring in the body as a result of this emotion. You may notice, for instance, that your breathing or your heart rate has changed. You may also notice temperature changes (feeling hot or cold) or pressure changes (feeling heavy or light). If you'd like, scan your entire body, becoming aware of how different regions of the body react when this emotion is felt. Here are some common experiences you can check in with yourself about as you scan each area:

    a. **Temperature changes:** Different areas feel hotter or colder

    b. **Pressure changes:** Different areas feel heavy, light, or compressed

    c. **Texture changes:** Different areas feel soft, sharp, or rough

    d. **Movement changes:** Different areas feel as though energy is moving them up or down (for instance, the sense that the stomach is dropping)

5. **Map out the sensations.** Using the following worksheet, map out the physical sensations that you notice as you focus on these thoughts and emotions. Write down the sensations on the right side of the worksheet, on the lines, and draw arrows to the part(s) of the body you experience those sensations in.

Physical Profile of Fear

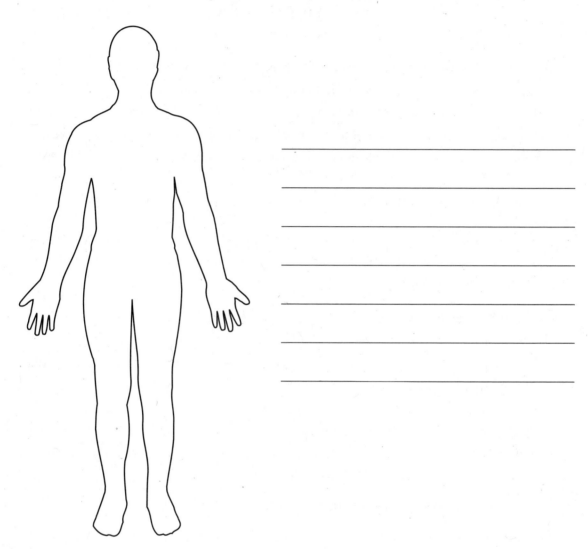

TIPS FOR CREATING A PHYSICAL PROFILE OF FEAR

Complete this exercise for any fear that causes you distress or interferes with you living life in the way you'd like. Thus, you may repeat this exercise several times.

When you are able to establish a connection between your thoughts, emotions, and physical sensations, you can understand why you might be avoiding certain experiences. Although it may seem that you avoid elevators because you fear becoming trapped, it's usually the distressing physical sensations, not the thoughts themselves, that lead you to avoid that situation. This is because distress in the body is often the core experience you are trying to avoid. Thus, knowing about and being able to work with these sensations can help you address the cause of your avoidance. Tolerating and managing these sensations will allow you to face feared situations with more confidence.

Skill 2: Reducing Distress with Bilateral Tapping (Bottom-Up and Top-Down)

Some types of psychotherapy incorporate bilateral movement (when you alternate movement between the left and right sides of your body), such as eye movements or bilateral tapping, as these can help clients quickly process distressing memories, thoughts, or fears. While it may seem strange to move your eyes (or some other area of the body) as you think about something upsetting, bilateral movement has actually been shown to help people desensitize or habituate to distressing thoughts. One reason for this may be that when you engage in bilateral movement while concentrating on something distressing, the movement slightly distracts you from fully focusing on the upsetting thoughts, which leads to faster desensitization to the distressing memory. Thus, incorporating bilateral movement into imaginal exposure techniques (where you imagine a feared situation, for instance) is a quick and easy way to reduce the distress you feel about feared situations.

1. **Identify an avoided person, place, or situation.** If you'd like, you can bring to mind the same avoided situation as you used in skill 1, or you can identify a different situation.

2. **Begin tapping your shoulders.** Gently tap your shoulders in sequence, tapping your left shoulder, then right shoulder, then left shoulder, etc., with each hand. You

may choose to tap each shoulder with the hand of that side or cross your arms and tap the opposite shoulder. As you do this, speed up your tapping a bit, spreading out each tap by about half of a second. It's important for this exercise for the tapping to be *fast*, not slow, in order to help distract you a bit from the distressing thoughts. Establish a fast rhythm with your tapping before moving on to the next step.

3. **Connect with the feared or avoided situation.** Close your eyes, envisioning the feared situation as clearly as you can. What are you most afraid of happening? Imagine that your worst fears about this situation are coming true. Where are you? What are you doing? Who else is there? What happens? Play this out in your mind as you'd imagine it happening in real time, still tapping your shoulders.

4. **Add deep breathing.** Staying here for a few moments, tapping and imagining the feared situation, begin to lengthen your breath a bit, taking longer inhales and exhales. Keep breathing, tapping, and imagining the situation for about five minutes.

TIPS FOR REDUCING DISTRESS WITH BILATERAL TAPPING

To significantly desensitize to thoughts about feared situations, you'll need to practice this technique several times. Try to practice every day for about two weeks, for five to ten minutes each day. Once you feel less distressed when you *think about* an avoided situation, it will become easier to actually *reenter* that situation. Although you'll still likely feel anxious when you go into that situation, that anxiety may feel more tolerable and less intense than if you had not practiced this technique.

If you begin to feel overwhelmed, or *too activated* when practicing this technique, open your eyes, take a break for a few minutes, and try standing up, walking around, or going outside for some fresh air. If you don't want to completely disengage with the feared situation, you may also imagine it with your eyes open, imagine just one part of it, or imagine one experience associated with it (such as one image or one moment in time, as opposed to the entire feared situation), to water it down a bit.

Skill 3: Facing Your Fears (Top-Down and Behavioral)

In the last two skills you were asked to bring to mind a feared, avoided situation, person, or place, to understand your physical and emotional reactions to it and begin to desensitize to it. These are great skills to practice to address your avoidance and agoraphobia. However, over time most people begin avoiding *multiple* or even *several* situations. Oftentimes, these situations have some theme in common. For instance, they all may involve crowds of people, being in small places, or a type of person (all men or all women, for example). Not all of these situations will elicit the same amount of anxiety; perhaps some of them cause slight discomfort, while others may incite terror.

In this exercise, you'll be asked to think about and identify as many feared situations, places, or people you can and list them on a fear hierarchy. The fear hierarchy, which appears in the Facing Your Fears worksheet, is a way of identifying feared and avoided situations and assigning to each situation a number that represents the level of anxiety you feel when you are faced with that situation. See the following example to better understand how this works.

Facing Your Fears Worksheet Example

On the fear hierarchy, list the least anxiety-provoking situations lower on the list (using the lines on the right) and the more anxiety-provoking situations higher on the list. For each anxiety-provoking situation, assign a distress score, from 1 to 100, where 1 represents no anxiety at all and 100 is the most anxiety you could imagine experiencing. Put the situation with the lowest distress score at the bottom of the hierarchy and the most distressing situation at the top.

90	Driving over bridge at rush hour (85)
80	
70	Getting on a crowded elevator (68)
60	Black Friday shopping at the mall (60)
50	
40	Standing in a long line at the store (40)
30	Being in a non–crowded elevator (27)
20	Driving in the car (23)
10	Going into the crawl space under house (10)

In this example, the theme of the avoided situations would likely be cramped spaces or situations where escape would be difficult. In these situations, this person likely feels trapped, and so they have started avoiding these types of situations (to some extent). Note that the person's distress score is listed to the right of each description, and the lower-anxiety situations are listed at the bottom of the hierarchy, whereas the more anxiety-provoking situations appear higher on the list.

As you fill out your own worksheet, think about situations that you've come to fear and have started avoiding since the trauma. Some of these may be situations you still engage with, but with hesitation or trepidation. In other words, it won't likely be the case that you completely avoid *all* of the situations you list, especially those appearing lower on the list. You can also visit http://www.newharbinger.com/48848 to download a copy of this exercise.

Facing Your Fears Worksheet

On the fear hierarchy, list the least anxiety-provoking situations lower on the list (using the lines on the right) and the more anxiety-provoking situations higher on the list. For each anxiety-provoking situation, assign a distress score, from 1 to 100, where 1 represents no anxiety at all and 100 is the most anxiety you could imagine experiencing. Put the situation with the lowest distress score at the bottom of the hierarchy and the most distressing situation at the top..

90

80

70

60

50

40

30

20

10

Now that you've completed the Facing Your Fears worksheet, the next step is to come up with a plan (the Facing Your Fears Plan) for how to increase your engagement, or participation, in those situations you listed. This is the step that makes most people even more anxious! However, this doesn't have to be intolerable. The strategy, broadly, will be for you to start very low on the fear hierarchy and tackle *those* situations before moving up the hierarchy to more intensely feared situations. While many people find it helpful, or even necessary, to seek a psychotherapist's help to face strongly feared or avoided situations, there are likely some gains you can make on your own.

Facing Your Fears Plan

Identify a feared situation that you experience at a distress level of 10 to 20. What is the situation you'd like to focus on?

_____

Repeat skills 1 and 2 with the feared situation. For example, if you listed "going to the mall on a quiet Sunday morning" as a 20 on your fear hierarchy, briefly repeat skills 1 and 2 from this chapter (Creating a Physical Profile of Fear and Reducing Distress with Bilateral Tapping), where you imagine going to the mall on a quiet Sunday morning. Revisiting these skills first will help you build confidence to engage in this activity in real life. Jot down any notes about what it was like as you engaged in these imaginal exercises.

_____

_____

_____

Schedule a time to directly engage with the feared situation in the near future. Expect that you'll still likely feel some discomfort, even though you've desensitized to the thought, to some degree. When are you going to engage in the activity?

_____

Engage in the feared situation. Remain in the feared situation until one of two things happens: your distress level has substantially decreased (by 50 percent), or you've stayed in the situation for forty-five minutes (because that's how long it might take for your body and brain to begin to relax and for your anxiety to decrease naturally). Try to refrain from engaging in distracting activities. While it might be tempting to call a friend, close your eyes (depending on the situation), or otherwise avoid parts of your experience, this is not recommended. (Distraction strategies _can_ be useful during some imaginal exposure exercises, and during trauma processing at certain times, but this is not recommended for real-life, "in vivo" exposure exercises.)

Reflect on your experience. How did you feel? Was it what you expected? Did you feel worse than you thought? Better? Do you want to make any changes for the next time you engage in the activity?

_____

_____

_____

_____

Set up a time to try the activity again. The more you can reengage in a feared situation, the easier it becomes. It may take several "exposures" to the feared situation for you to feel more at ease, but you'll find that even when you feel anxious in that situation, it becomes more bearable over time and your confidence will increase. When do you plan on practicing this again? Jot down your schedule below.

_____

_____

TIPS FOR FACING YOUR FEARS

It cannot be overstated that in order to make gains, you need to "practice, practice, practice!" Brain change occurs as a result of repetition and consistency, so the best way to tolerate and reduce anxiety related to feared situations is to repeatedly engage in them, using the steps outlined in this chapter. Although some feared situations, such as flying on a plane, may not be things you can practice daily, or even weekly, you can identify and consistently practice avoided situations that you can regularly engage with.

Skill 4: Overcoming Fear-Based Thinking (Top-Down)

A main reason you might develop avoidance and agoraphobia is that your body starts to send you danger signals when you encounter situations that seem to have some element in common with what you experienced during trauma. This process often occurs unconsciously, which is why somatic awareness and behavioral exposure techniques are important. However, over time your mind may also begin to play a large role in your avoidance. For example, your mind may start telling you that if you drive on the highway, you'll become trapped or have a panic attack. Or your mind might tell you that being in a crowd of people is not safe. When these thoughts occur, they exacerbate the anxiety, even if the thoughts are irrational or untrue. This technique helps you address these thoughts about feared and avoided situations so that you can more easily reengage with them. Use the worksheet following the directions to explore your fear-based thinking. You can also visit http://www.newharbinger.com/48848 to download a copy of this exercise so you can work on overcoming different fears from your hierarchy.

1. **Name the fear.** Begin by thinking about the feared situation you identified in skills 1 and 2 (Creating a Physical Profile of Fear and Reducing Distress with Bilateral Tapping, respectively), or any of the situations you listed in your Facing Your Fears worksheet. Begin with an example from the lower end of your hierarchy. Bringing that situation to mind, ask yourself, "What am I afraid of happening in this situation?" This will clarify what your fear is with regard to that situation. Your mind may tell you that you're not safe in that situation, that something bad could happen, or that someone is not trustworthy and might hurt you.

2. **State why the situation is important to you.** You've identified something you fear and perhaps that you've been avoiding to some extent. Now ask yourself, "Why bother working on this issue at all? What makes this feared situation important?" If it didn't matter to you, you wouldn't be motivated to change. For example, if you've been avoiding movie theaters, maybe you want to be able to go see movies with your children, if they love movies.

3. **Identify the underlying value.** When something is important to us, there's usually an underlying value that it fulfills. Values may include service to others, integrity, success, family, faith, financial well-being, parenting, hard work, and so on. Thinking about why this feared situation is important to you, ask yourself, "What does this reflect about my values? What value does it fulfill if I can engage in this feared situation?" In the example of going to the movie theater with your children, it could be that the value you fulfill by taking them to the movies is quality time, parenting, or family.

4. **Imagine success.** You already know that your mind can tell you all sorts of things about why a feared situation could be dangerous, and you've probably played out a lot of potential catastrophic outcomes. However, have you considered what might go *well*? Or the success you might feel if you were able to do it? Take a few moments to identify how the situation might go well, what the upsides of engaging in it might be, and how you'll feel if you face it. Also consider how it might impact others if you face this fear. In the movie theater example, you might envision how your children will react when you take them to the movies.

5. **Rehearse positive outcomes.** With the positive thoughts still in your mind, write down, as a story, how this situation could go really well. To do this, imagine yourself right before going into the situation, and how you'd like to feel. Then imagine engaging in the situation and envision having the best possible experience. The reason this step is important is because your brain has likely rehearsed the terrible outcomes of this situation, and these distressing thoughts may be worsening your fear. Rehearsing how being in the situation might go well is a way of preparing your brain to expect the best!

Overcoming Fear-Based Thinking Worksheet

My fear: _____

Why overcoming this fear is important to me: _____

_____

_____

The underlying value: _____

_____

What success would look like: _____

_____

_____

The most ideal experience I can imagine: _____

_____

_____

_____

_____

_____

_____

TIPS FOR OVERCOMING FEAR-BASED THINKING:

Consider practicing this skill immediately after you practice skill 2, Reducing Distress with Bilateral Tapping. Imagining and desensitizing to your worst fear about a situation, followed by imagining the best-case scenario with that situation, can help you more deeply desensitize to your anxious thoughts about the situation. Also practice this skill in isolation, as a meditative practice, both first thing in the morning and at night before going to sleep. Shifting toward positive thoughts at the beginning and end of the day can help you start and end your days on a high note.

A Pause to Reflect

This chapter has focused on ways to combat avoidance and agoraphobia, which most people find to be quite difficult! Congratulate yourself on being open to engaging in some very challenging work, and remember to seek professional guidance from a psychotherapist if you feel stuck. As you reflect on what you've learned here, think about the skills that you've found helpful. These might include skills described in this chapter as well as other techniques you've read about or learned about in therapy or other contexts. The following worksheet can help you organize your thoughts by listing the exercises that you've either found helpful or would like to try in the future. You can also visit http://www.newharbinger.com/48848 to download a copy of this exercise.

Skills for Managing Avoidance, Isolation, and Agoraphobia

The following chart is a reminder of the skills in this chapter and a place to record how each skill works for you. When you notice feeling anxious about engaging in certain situations, and it is negatively impacting your life, choose a skill that is appropriate for the situation. The second column tells you when the best time to practice is. Over the next week, make a note of when and how you try the skill in the third column. Then, take a moment to reflect on your experience. If you have learned other useful skills or techniques that help, add them to the chart in the blank spaces. If you need more space, you can print more pages or keep a journal to reflect on your experiences.

Skill/Technique	Best Time to Practice	When I Practiced the Skill (Day and Time as Well as Description of the Situation)	Reflect on the Experience. Was the Skill Useful?
Creating a Physical Profile of Fear	When I am feeling afraid and want to know more about this experience in my body.		
Reducing Distress with Bilateral Tapping	When I'm feeling upset or triggered and want to feel better.		
Facing Your Fears	When I want to practice facing my fears in a structured manner.		
Overcoming Fear-Based Thinking	When I want to tackle distressing thoughts.		

CHAPTER 9

Sleeping Difficulties

Eva lost her home in a tornado last year and says her main difficulty since then is that she can't sleep. She says it's been almost a year since she slept a full night, and she is fed up with being chronically exhausted. Although some over-the-counter medications used to help her fall asleep, they never helped her stay asleep. Moreover, these over-the-counter medications no longer help her fall asleep. Eva says she's increasingly hopeless about ever feeling rested again and is frustrated about the "racing thoughts" she can't seem to stop having when it's time for bed.

Check-In: Your Experiences

After reading about Eva, take a moment to think about your experiences. You may choose to reflect and journal on the following questions:

- Do you have difficulties falling or staying asleep at night? What are the difficulties?

- When did your sleep difficulties begin? Do they seem to be trauma-related or do they predate any trauma?

- About how many hours total do you sleep per night? How many would you like to be sleeping to feel rested?

How Sleep Difficulties May Impact Your Life

Sleep difficulties or insomnia can arise in a few different ways. When mental health providers ask about insomnia, they tend to ask about the following four types:

1. **Difficulty falling asleep.** This is a common type of insomnia where you cannot initially fall asleep. While everyone experiences this from time to time, with PTSD it may take an hour or more for you to fall asleep. Sometimes, when you anticipate having difficulty falling asleep, you may become so anxious about it that the anxiety itself becomes a barrier to sleep!

2. **Difficulty staying asleep.** Often, those with PTSD wake up in the middle of the night, sometimes multiple times. While this may not be bothersome to you, it becomes a problem if you can't quickly fall back asleep.

3. **Waking up too early.** Strongly associated with some types of depression, this form of insomnia also sometimes affects those with PTSD. In this type of insomnia, you consistently wake up substantially earlier than you would like to and can't go back to sleep.

4. **Reduced sleep quality.** This type of insomnia is a bit tricky to identify sometimes, especially when combined with one of the other types of insomnia. It occurs when you get adequate (or near adequate) sleep, perhaps seven to eight hours or more in one night, but still feel tired and worn out the next day. This is often due to poor sleep quality. You might feel like you've hardly sleep at all even though you know you did sleep.

If one or more of these sleep issues has persisted for two or more weeks, seek professional help. This is because insomnia can significantly worsen, or even contribute to the development of, many different mental health and medical conditions. Also, if you are currently in psychotherapy for mental health–related concerns, you may find it less helpful, and may notice your progress slowing, if you are also dealing with chronic sleep issues. When sleep diminishes, so does quality of life.

The following brief self-assessment can help you determine whether sleep difficulties are particularly problematic for you and might require intervention. You can also visit http://www.newharbinger.com/48848 to download a copy of this exercise.

Sleep Difficulties Self-Assessment

Read the following and circle the best response from 0 to 3, where 0 = None/never, 1 = A little bit/sometimes, 2 = A moderate amount/often, and 3 = A lot/most of the time:

I have a hard time going to sleep.	0	1	2	3
I have a hard time staying asleep at night.	0	1	2	3
I wake up too early in the morning.	0	1	2	3
I do not feel rested after sleeping at night.	0	1	2	3
I sleep fewer hours than I need to be rested.	0	1	2	3
I sleep less than six hours per night.	0	1	2	3
My sleep difficulties have lasted more than one month.	0	1	2	3
My sleep difficulties started or have worsened after a traumatic event.	0	1	2	3
I am anxious about going to bed because I am afraid I won't be able to fall asleep or that I might have nightmares.	0	1	2	3
I have nightmares.	0	1	2	3
I wake up feeling panicked in the morning.	0	1	2	3
I am unhappy with my sleep patterns.	0	1	2	3
My sleep difficulties interfere with my daily functioning (affect concentration, work functioning, memory, etc.).	0	1	2	3
My sleep difficulties distress or worry me.	0	1	2	3
I have to try hard to sleep.	0	1	2	3
I take more than thirty minutes to fall asleep.	0	1	2	3
I sleep more than ten hours per night.	0	1	2	3
My sleep is fitful or restless.	0	1	2	3
Total				

Add up the total score from the above items, and record them next to "Total." A score of 20 or higher indicates you may benefit from working on your sleep difficulties.

Use the following exercise to reflect on your self-assessment score. You can also visit http://www.newharbinger.com/48848 to download a copy of this exercise.

Reflect on Your Self-Assessment Score

What did you score on the self-assessment? Did you encounter any surprises? Take a moment to process what you learned in your self-assessment by considering the following questions.

Did you learn anything new about your relationship with sleep? Oftentimes, people know they aren't sleeping well, but they don't realize how nervous or worried they feel at bedtime. What is your relationship with bedtime and sleep like?

_____

_____

_____

_____

Were there items on the assessment you hadn't thought about before now?

_____

_____

Were there any sleep-related issues you didn't realize were so impactful for you?

_____

_____

_____

Why Sleep Difficulties After Trauma Are So Common

Approximately 80 percent of people with PTSD report experiencing sleep difficulties, and the percentages are similar for those experiencing depression or anxiety disorders. In fact, insomnia is a key symptom of *many* mental health conditions! There are several causes of insomnia, which is one reason sleep difficulties are so common. Most of the time, the main issue is that the amygdala (the "smoke alarm" of the brain) is activating, making you feel revved up, anxious, or on guard, and this makes it difficult to sleep well. Specific possible reasons for your insomnia may include one or more of the following.

1. **Nightmares.** Nightmares are a common symptom of PTSD and can make it difficult to stay asleep. When nightmares wake you up, you may experience strong anxiety or panic symptoms, which are not conducive to sleep. When your stress response is active in your brain and body and threat signals are present, it makes sense that sleeping would be difficult or impossible, as it is not adaptive to your survival to sleep if there is a danger present. Even if you're safe, when your brain and body tell you otherwise, sleep will not come easily. Also, recurring nightmares can lead to anticipatory anxiety about falling asleep, where you fear going to bed at night due to the possibility of experiencing a nightmare.

2. **Racing or upsetting thoughts.** This is arguably the most common reason people have a hard time getting to sleep or staying asleep. If you are experiencing depression, anxiety, or PTSD, you may find it difficult to "shut off" your mind when it's time for bed. If you have depression, you may ruminate about things that have happened in the past; whereas if you have anxiety, you may worry about the future. If you have PTSD, you may experience both—recalling terrible things that happened in the past and also feeling anxious about the future.

3. **Distressing physical sensations.** As described in chapter 4, distressing physical sensations are a common experience after trauma, especially when you are faced with traumatic reminders. These sensations may be frequent and prolonged, and can arise spontaneously even when you can't identify any trauma triggers. For example, you may lie down to go to sleep at night and immediately notice your heart racing and muscles tensing. These types of sensations, which reflect the stress response, will likely make it difficult to fall asleep.

4. **Not feeling safe.** Even if you know, logically, that you are likely safe at night, you may still feel unsafe if you've experienced trauma. (See chapter 6, Issues Around Safety and Control, for more information on this topic.) This is because sleep can make you feel vulnerable. When you're asleep you can't defend yourself, and you could be caught off guard. If you were traumatized at night, sleep can be particularly anxiety-provoking.

In addition, insomnia may be related to medical conditions or a misalignment of your sleep drives. A detailed discussion of these topics is beyond the scope of this book, but if your insomnia has persisted for more than a couple of weeks, meet with your physician to explore possible medical issues that might be contributing to or causing the insomnia. Thyroid-related issues, sleep apnea, chronic pain, and other conditions can interfere with sleep and need to be addressed by a medical professional.

While sleep difficulties are extremely common after trauma (and otherwise!), not every person's specific challenges look exactly alike. Some people have problems falling asleep, while for others, staying asleep is the main issue. For yet others, it's nightmares that interfere with their sleep. Complete the following exercise to better understand your own sleep difficulties, referring back to the self-assessment earlier in this chapter if needed. You can also visit http://www.newharbinger.com/48848 to download a copy of this exercise.

Your Sleep Difficulties

What's your biggest sleep complaint: falling asleep, staying asleep, early awakenings, or something else?

_____

What seems to be the cause of your main sleep difficulty? For instance, you can't "turn off your thoughts" or there are distractions in your environment that make it hard to sleep.

_____

_____

What have you tried so far for your sleep problems? What has been helpful, if anything? What tends to make the problems worse?

_____

_____

Terms Your Therapist Might Use to Explain This Challenge

If your experiences are similar to Eva's, your therapist may use the following descriptors and terms in your discussions, or in your therapy notes:

- **Arousal and reactivity symptoms.** Sleep impairment is explicitly noted as a symptom in many mental health conditions, including PTSD and other trauma- or stressor-related disorders. In the *DSM-5*, sleep difficulties are conceptualized as an arousal and reactivity symptom of PTSD, along with other symptoms such as hypervigilance, poor concentration, and irritability. Insomnia is considered a hyperarousal symptom because those suffering from post-trauma symptoms often find it difficult to relax enough to sleep well.

- **Reexperiencing symptoms.** While sleep impairment is usually thought of as a symptom of hyperarousal, if the cause of the insomnia is nightmares, your therapist may refer to it as a reexperiencing symptom. This is because nightmares are considered a way that people relive aspects of a traumatic event. While the nightmares may not be obvious reenactments of the traumatic event, they often have elements in common with the traumatic incident.

What Your Brain Needs

When insomnia occurs, several things may happen in the brain, as sleep (and therefore, insomnia) is quite complex! However, a couple of specific brain areas have been shown to be involved in insomnia. These are also brain regions that tend to become dysregulated after trauma. First, hyperactivation of the brain's "smoke alarm" (amygdala) can contribute to insomnia, as it sends threat signals throughout the brain and body that make it difficult to fall and stay asleep. Second, during insomnia the "thinking area" of the brain (prefrontal cortex) becomes dysregulated, which is associated with worry, rumination, and an inability to control anxiety and stress. Here is what your brain needs if you're experiencing insomnia.

1. Bottom-up techniques that reduce action of the "smoke detector" of the brain (amygdala). Relaxation techniques that turn down the smoke detector of the brain can promote sleep. In addition, basic behavioral changes, such as consuming less caffeine or refraining from watching horror movies before bed, can help calm the brain's smoke detector.

2. Top-down techniques that regulate the "thinking center" of the brain (prefrontal cortex). Cognitive techniques—skills that use your prefrontal cortex and require intentional thought and effort—can be helpful if you've noticed your insomnia is related to thoughts that seem to spin out of control. Learning how to catch these thoughts, address them (if needed), and redirect your attention toward something more calming is both important and possible with some skill building.

The next section details four skills that can help you learn how to calm the brain's smoke alarm and regulate the thinking center of the brain, allowing you to sleep better.

Skills for Meeting the Challenge

This section focuses on four exercises that can help you learn more about, and better manage, your sleep difficulties.

Skill 1: Sleep Log (Top-Down)

To better understand possible causes and contributing factors to your insomnia, it can be helpful to keep a sleep log. While it may be daunting to answer so many questions about your experiences on a daily basis, the sleep log only takes a few minutes to complete, and it can provide invaluable information about why you might be experiencing sleep difficulties. You can also visit http://www.newharbinger.com/48848 to download a copy of this exercise so you can complete it as often as needed.

Sleep Log

Date	
What time did you go to bed?	
Did you do anything in bed (scroll social media, read a book, watch TV) before trying to go to sleep?	
About when did you fall asleep?	
How many times did you wake up and when?	
How long were you awake each time?	
What woke you up each time?	
What kept you awake, if anything?	
When did you wake up for good in the morning?	
What time did you get out of bed?	

How many hours did you sleep in total?	
Did you take any naps on this day? When and for how long?	
Did you take any medications before bed?	
Did you eat or drink within three hours of bed? What?	
Did you consume caffeine on this day?	
Who was in bed with you on this night (pets, spouse, children)?	
What activities did you engage in within three hours of bedtime?	
On a scale of 1 to 10, where 1 is no stress at all and 10 is the most stress you've ever felt, how stressful was this day overall?	
Other comments and information:	

TIPS FOR SLEEP LOG

Maintain a sleep log for at least two weeks to fully understand your sleep patterns. After a couple of weeks of keeping a sleep log, you can begin to see patterns emerging, including thoughts, emotions, situations, foods, medications, and activities that could be impacting your insomnia. For example, you may discover that your insomnia worsens if you consume caffeine after 12 p.m., sleep with your cat, watch a scary movie before bed, or work on your laptop at night. Once you can identify the factors that contribute to your insomnia, you can begin to address them directly.

Skill 2: Creating a Nighttime Ritual (Bottom-Up and Top-Down)

Becoming mindful about how you spend the last couple of hours before bedtime can help you tackle difficulties falling asleep. Ideally, you will create your own personal nighttime ritual based on your lifestyle, preferences, and tendencies. To help you get started, here are several recommended activities and tips for you to consider as you decide what you want your own ritual to look like. After reviewing the recommendations, you can draft your own nighttime ritual on the following worksheet. You can also visit http://www.newharbinger.com/48848 to download a copy of this worksheet.

NIGHTTIME RITUAL RECOMMENDATIONS

Suggested activities two hours before bedtime:

- Finish up low-stress work activities.

- Put children to bed, lock up your home, put pets outside for the last time.

- Wind down social media engagement.

- Wrap up serious conversations with your partner.

- As needed, make plans for the following day.

- If needed, have a snack that is high in tryptophan (to induce sleepiness).

Suggested activities one hour before bedtime:

- Take a hot shower or bath (your body will expend energy trying to heat up when it experiences the colder air, which can make you sleepy).

- Drink decaffeinated tea or warm milk.

- Ensure your bedroom is not too hot or cold.

- Begin dimming lights in your environment.

- Disconnect from screens.

- Engage in a repetitive activity, such as knitting or sewing.

- Practice meditation or a mindfulness or relaxation exercise.

- Write down anything you think you need to remember for the following day, so that you do not have to think about it while in bed.

- Write down any problems that have been on your mind and put the list where you can review it the next day. If the problems begin to intrude in your mind, remind yourself that they are already accounted for and you plan to address them again tomorrow.

Suggested activities to do while in bed:

- Read a calming book with dim (but adequate) lighting.

- Complete a gratitude journaling exercise.

- Enjoy sex with your partner or alone.

- Cuddle with and pat your pets.

- Listen to calming music.

- Pray or meditate.

- Practice deep breathing, autogenic training, or another relaxation exercise.

Suggested daytime behavior modifications:

- Limit caffeine to morning hours and, if possible, consider eliminating caffeine.

- Limit or avoid foods that are associated with heartburn.

- Complete rigorous exercise by late afternoon.

- Get access to sunshine or bright light within ten minutes of waking up in the morning.

- If possible, set a consistent wake-up time each morning and adhere to it.

- Avoid daytime napping.

Nighttime Ritual Worksheet

Daytime behavior modifications I'd like to try to help my insomnia:

_____

_____

_____

Things I'd like to do two hours before bed:

_____

_____

_____

Things I'd like to do one hour before bed:

_____

_____

_____

Things I'd like to do while in bed:

_____

_____

_____

TIPS FOR CREATING A NIGHTTIME RITUAL

Update this plan every month or so, as you learn what works for you over time. You may find that what you thought was working isn't, or that something needs to be altered in your plan.

Skill 3: Creating an Early Awakening Plan (Top-Down)

If you have issues waking up too early, or can't seem to stay asleep during the middle of the night, you may want to consider making an early awakening plan for how to handle unwanted awakenings. When you can anticipate this issue and have a plan in mind for how to handle it, it can make awakenings feel less upsetting and disruptive. After reviewing the recommendations, you can draft your own early awakening plan on the following worksheet. You can also visit http://www.newharbinger.com/48848 to download a copy of this worksheet.

EARLY AWAKENING RECOMMENDATIONS

Ways to handle waking up at night or too early in the morning:

- When you first wake up, note what woke up you, if possible. If something in your environment needs to be addressed (a snoring partner, for instance), attempt to address it quickly.

- Practice a short, simple, relaxation technique such as deep breathing or progressive muscle relaxation (these techniques are described in other chapters of this workbook).

- Follow the "twenty-minute rule": If you wake up and do not fall back asleep within about twenty minutes, get out of bed and leave the bedroom. Relocate to another room and, keeping the lights dim if possible, engage in a repetitive activity (such as crocheting) or read a book you find to be boring. Only return to bed when you feel sleepy and are having a difficult time staying awake to engage in the activity you selected.

- Write down any thoughts or problems that seem to be keeping you awake, and put the piece of paper on your bedside table. When one of those thoughts reenters your mind, remind yourself that you do not need to think about it right now and can revisit the topic the next day.

Ways to handle nightmares that wake you up:

- If needed, turn on a dim light and reorient yourself to your bedroom. Take a look around and notice what you see, reminding yourself of where you are and that you are safe.

- When you wake up, practice a short, simple, relaxation technique such as deep breathing or progressive muscle relaxation. This can calm your body's stress response and prepare you for sleep.

- Reconnect with your pet(s), holding and cuddling them.

- Keep a night-light on if it helps you feel oriented, grounded, and safe.

- If you are practicing nightmare rescripting, take a moment to rewrite the nightmare before lying back down to go back to sleep (see the next skill, Nightmare Rescripting, for specific instructions).

Early Awakening Worksheet

What I'm going to do if I wake up before it's time:

_____

_____

_____

What I'm going to do if a nightmare wakes me up:

_____

_____

_____

Things I'd like to do while in bed if I wake up:

_____

_____

_____

Things I'm going to do if I stay awake more than twenty minutes after an early awakening:

_____

_____

_____

TIPS FOR CREATING AN EARLY AWAKENING PLAN

Update this plan as needed. Keep this workbook or your worksheet near your bed, to reference when you wake up. That way, you will not forget what to try in those moments, when you might be feeling drowsy and lack clarity of mind.

Skill 4: Nightmare Rescripting (Top-Down)

Because what we do and think about during the day can impact how we dream at night, it is possible to influence dreams, including nightmares. One technique that is taught in multiple types of psychotherapies is called "nightmare rescripting," and it has been shown to reduce nightmare occurrence and intensity. This exercise might be of benefit to you if:

- Your nightmares tend to be similar, or at least have common elements or themes

- Your nightmares occur frequently and disrupt your sleep

- You remember your nightmares fairly well

An easy-to-follow version of the nightmare rescripting technique appears in the following worksheet. You can also visit http://www.newharbinger.com/48848 to download a copy of this exercise.

Nightmare Rescripting Worksheet

Identify the common elements or themes in the nightmare that you'd like to rescript.

_____

_____

_____

Write out the nightmare in its entirety. If the nightmare is a bit different each time, write out an example of how the nightmare *might happen*, being sure to include the common elements and themes you identified in step 1. Specifically, as you write out your nightmare, try your best to make it a story with a beginning, middle, and ending. While most nightmares do not usually have a conclusion or clear ending, there is often a point at which it suddenly ends; be sure to note when this occurs and what happens immediately before the nightmare ends.

_____

_____

_____

_____

_____

If desired, identify a couple of elements of the nightmare you'd like to change. It is important that you do not change *too much* about the nightmare or the rescripting may not work as well. Choose one to three small details or elements of the nightmare to change, and write them down below. For example, if in the nightmare a person has a knife, you might change the knife to a zucchini to make it less threatening.

_____

_____

_____

Create an ending to the nightmare. As noted earlier, most nightmares drop off with no good conclusion or resolution. To rescript the nightmare, it needs to have a positive ending. Write out a possible ending to the nightmare that would reduce the distress you feel about it. It is not important that this ending be realistic; as long as it makes the nightmare less scary and resonates with you, it is fine.

_____

_____

_____

_____

Incorporating the changes and additions you've made to the nightmare, rewrite the nightmare in its entirety below. Remember, the nightmare should have a beginning, middle, and ending that is not distressing. Also, a few small, upsetting pieces of the nightmare should be replaced with neutral or positive elements.

_____

_____

_____

_____

_____

TIPS FOR NIGHTMARE RESCRIPTING

After completing this exercise, rewrite the modified, completed version of the nightmare (as written in step 5) once per week. When you do this, be sure to rewrite it verbatim. Additionally, consider reading the new version of the nightmare each night before bed, repeatedly, for about five minutes. Rehearsing new, less distressing versions of nightmares has been shown to reduce the frequency and intensity of old, incomplete nightmares. After a while, you may even notice that when the nightmare occurs, the new version is what plays out!

Keep in mind that you can do nightmare rescripting for *multiple* nightmares. Sometimes people have three or four recurring nightmares and rescript all of them. You may elect to keep a sleep notebook, where you keep daily sleep logs, updated nighttime rituals, early awakening plans, and nightmare rescripting notes.

A Pause to Reflect

Now that you've come to the end of this chapter, check in with yourself about what has been helpful so far as you work on improving your sleep. The exercises and tools may be from this chapter and also from other chapters (especially relaxation-type exercises such as deep breathing). The following worksheet can help you organize your thoughts by listing the exercises that you've either found helpful or would like to try in the future. You can also visit http://www.newharbinger.com/48848 to download a copy of this exercise.

Skills for Managing Sleep Difficulties

The following chart is a reminder of the skills in this chapter and a place to record how each skill works for you. When you notice sleep difficulties are negatively impacting your life, choose a skill that is appropriate for the situation. The second column tells you when the best time to practice is. In the third column, make a note of when and how you try the skill over the coming week. Then, take a moment to reflect on your experience. If you have learned other useful skills or techniques that help, add them to the chart in the blank spaces. If you need more space, you can print more pages or keep a journal to reflect on your experiences.

Skill/Technique	Best Time to Practice	When I Practiced the Skill (Day and Time as Well as Description of the Situation)	Reflect on the Experience. Was the Skill Useful?
Sleep Log	Every day for at least two weeks.		
Creating a Nighttime Ritual	When I want to try new ways to sleep better by creating a ritual.		
Creating an Early Awakening Plan	When I want to try new ways to address early awakenings.		
Nightmare Rescripting	When I have recurring nightmares that cause me distress.		

Concluding Thoughts

You may find that you work through various chapters of this workbook multiple times, tackling the challenges in slightly different ways over time. Thus, there may not be a clearly defined end point as you work toward recovering from trauma, and that's totally normal. In fact, most people find that their recovery is a work in progress and they learn more and more about themselves as they continue to journey inward! That said, it can be helpful to occasionally zoom out and check in with yourself, reflecting on what you've tried, what works or doesn't, what's improved, and what still needs your attention. Engage in this type of check-in once every few months, to keep track of how you're growing and shifting.

The following exercise aims to help you check in with your progress. You can also visit http://www.newharbinger.com/48848 to download a copy of this exercise.

Self-Assessment: Where Are You Now?

Take a few moments to reflect on what has changed for you since you last checked in. Organization isn't important; it's fine to treat this as a sort of "stream of consciousness." Also, it's okay if all of the changes aren't what you hoped they'd be, or positive. You may improve in some ways but slip a bit in other ways or in other domains.

What has changed for you: What do you notice right now as you reflect? How are you feeling, thinking, or behaving differently than you have in the past?

_____

_____

_____

_____

Skills that have helped: Think about what skills or techniques you've incorporated into your life since you last checked in. Are there ones that stick out to you as especially helpful? Which tools are working for you, that you'd like to continue using between now and the next time you check in with yourself?

_____

_____

_____

_____

_____

Symptoms that have improved: What exactly has changed? Get a bit more specific about what's different. For example, are you getting along better with others? Getting out more? Feeling more joy? Sleeping better? What's different? Following is a list of symptoms that a lot of trauma survivors struggle with. As you look at them, consider whether any of these have improved for you. Many of the symptoms refer to the challenges presented in the chapters of this book, but some are additional ones to consider. You may also add your own symptoms, as this list is not exhaustive.

- Anxiety

- Being on guard or hypervigilant

- Feeling angry or enraged

- Distressing physical sensations associated with stress (racing heart rate, nausea, jittery)

- Apathy or difficulty connecting to or expressing emotions

- Feeling unsafe

- High need for control

- Difficulty trusting others

- Tumultuous or tense relationships

- Unwanted or intrusive thoughts about the trauma(s)

- Frequent worry that is hard to manage

- Isolation or agoraphobia

- Insomnia or other sleep difficulties

- Concentration difficulties

- Loss of interest in things you used to value

- Feeling unclear about who you are or what you want

- Thoughts of self-harm

- Feelings of guilt or shame

- Feeling upset when you think about the trauma(s)

- Avoidance of people, places, situations, thoughts, or emotions that remind you of the trauma(s)

- Negative thinking about yourself, others, or the world

- Reckless or risky behavior

- _____

- _____

- _____

- _____

- _____

Symptoms to continue working on: Consider which symptoms you feel you most need to work on moving forward. What's front and center for you, causing you the most pain or difficulty in your life? What do you want to work on between now and your next check-in?

_____

_____

_____

_____

_____

When to Seek Professional Help

As you continue into the next phases of your trauma recovery, you may want to consider starting (or perhaps continuing!) therapy with a licensed psychologist or psychotherapist. For a lot of survivors, therapy can provide a major boost in their improvement and can teach skills that aren't possible to learn in a book. Also, the strong connection and sense of trust you can build with a therapist can be healing *in and of itself*. There's nothing that can truly replace the gift of therapy!

You may decide to pursue psychotherapy if you find that your progress has stalled or your symptoms are worsening. Seek professional help if you start having thoughts about hurting yourself or others or feel increasingly hopeless about getting better. Trained therapists can help you navigate these difficult feelings and explore ways to keep you safe and on the right track. While it can be hard to ask for help, working with a trusted professional can bring you the most transformational growth.

Acknowledgments

This book would not be possible without the ongoing help of my amazing contacts at New Harbinger, several colleagues who consulted with me during the writing process, and the relentless support of my husband, Tim. Thank you all so much!

References

American Psychiatric Association. 2013. *DSM-5*. Washington, DC: American Psychiatric Association.

Bandelow, B. 1995. "Assessing the Efficacy of Treatments for Panic Disorder and Agoraphobia. II. The Panic and Agoraphobia Scale." *International Clinical Psychopharmacology* 2: 73–81.

Beck, A. T., N. Epstein, G. Brown, and R. A. Steer. 1988. "An Inventory for Measuring Clinical Anxiety: Psychometric Properties." *Journal of Consulting and Clinical Psychology* 56: 893–897.

Deninger, M. 2021. *Multichannel Eye Movement Integration: The Brain Science Path to Easy and Effective PTSD Treatment*. Grand Rapids, MI: Gracie Publications.

Preece, D., R. Becerra, K. Robinson, J. Dandy, and A. Allan. 2018. "The Psychometric Assessment of Alexithymia: Development and Validation of the Perth Alexithymia Questionnaire." *Personality and Individual Differences* 132: 32–44.

Russell, D., L. A. Peplau, and M. L. Ferguson. 1978. "Developing a Measure of Loneliness." *Journal of Personality Assessment* 42: 290–294.

Snell Jr., W. E., S. Gum, R. L. Shuck, J. A. Mosley, and T. L. Kite. 1995. "The Clinical Anger Scale: Preliminary Reliability and Validity." *Journal of Clinical Psychology* 51 (2): 215–226.

Spielberger, C., R. Gorsuch, R. Lushene, P. Vagg, and G. Jacobs. 1983. *Manual for the State-Trait Anxiety Inventory*. Palo Alto, CA: Consulting Psychologists Press.

Sweeton, J. 2019. *Trauma Treatment Toolbox*. Eau Claire, WI: PESI Publishing and Media.

Taylor, G. J., D. Ryan, and R. M. Bagby. 1985. "Toward the Development of a New Self-Report Alexithymia Scale." *Psychotherapy and Psychosomatics* 44: 191–199.

Jennifer Sweeton, PsyD, is a clinical and forensic psychologist, author, and internationally recognized expert on trauma, post-traumatic stress disorder (PTSD), and the neuroscience of mental health. She is a sought-after expert who has trained more than 15,000 mental health professionals in all fifty US states and in over twenty countries. She is author of *Trauma Treatment Toolbox*. She resides in the greater Kansas City, MO area.

Real change *is* possible

For more than forty-five years, New Harbinger has published proven-effective self-help books and pioneering workbooks to help readers of all ages and backgrounds improve mental health and well-being, and achieve lasting personal growth. In addition, our spirituality books offer profound guidance for deepening awareness and cultivating healing, self-discovery, and fulfillment.

Founded by psychologist Matthew McKay and Patrick Fanning, New Harbinger is proud to be an independent, employee-owned company. Our books reflect our core values of integrity, innovation, commitment, sustainability, compassion, and trust. Written by leaders in the field and recommended by therapists worldwide, New Harbinger books are practical, accessible, and provide real tools for real change.

 **newharbinger**publications

MORE BOOKS from
NEW HARBINGER PUBLICATIONS

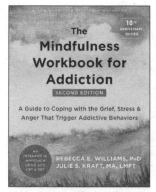
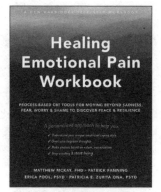

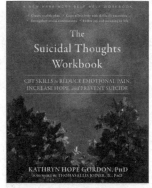

Did you know there are **free tools** you can download for this book?

Free tools are things like **worksheets**, **guided meditation exercises**, and **more** that will help you get the most out of your book.

You can download free tools for this book— whether you bought or borrowed it, in any format, from any source—from the New Harbinger website. All you need is a NewHarbinger.com account. Just use the URL provided in this book to view the free tools that are available for it. Then, click on the "download" button for the free tool you want, and follow the prompts that appear to log in to your NewHarbinger.com account and download the material.

You can also save the free tools for this book to your **Free Tools Library** so you can access them again anytime, just by logging in to your account! Just look for this button on the book's free tools page.

+ Save this to my free tools library

If you need help accessing or downloading free tools, visit **newharbinger.com/faq** or contact us at **customerservice@newharbinger.com.**